D0955812

Beyond Engineering

BY THE SAME AUTHOR

To Engineer Is Human:
The Role of Failure in
Successful Design

Beyond Engineering

Essays and Other Attempts to Figure Without Equations

HENRY PETROSKI

Illustrations by Karen Petroski

ST. MARTIN'S PRESS · New York

Design by Mina Greenstein

Some of this material has appeared, often in somewhat different form, in: *The Christian Science Monitor, The Chronicle of Higher Education, College English, Creative Computing, The Futurist, The Guide, Issues in Engineering —Journal of Professional Activities, Proceedings of the American Society of Civil Engineers, The New York Times, Politics Today, Prairie Schooner, The Reading Teacher, Science 80, Science 81, The South Atlantic Quarterly, Technology Review, The Virginia Quarterly Review, The Washington Post,* and *The Washington Post Magazine.*

Library of Congress Cataloging-in-Publication Data

Petroski, Henry.
Beyond engineering.

1. Engineering. I. Title
TA155.P48 1986 620 86–3759
ISBN 0-312-07785-8

First Edition
10 9 8 7 6 5 4 3 2 1

TO STEPHEN,
with whom I build bridges

Contents

Preface xi
1 · Introduction: Writing as Bridge-Building 1

At the Laboratory

2 · Of Two Libraries 17
3 · Amory Lovins Woos the Hard Technologists 24
4 · Soft Technology Is Hard 31
5 · The Gleaming Silver Bird and the Rusty Iron Horse 35
6 · Reflections on a New Engineer's Pad 39
7 · Logon Proceeding 46
8 · How Poetry Breeds Reactors 53

At the University

9 · You Can't Tell an Engineer by His Slide Rule 65
10 · A New Generation of Engineers 72
11 · The Quiet Radicals 79
12 · Apolitical Science and Asocial Engineering 84
13 · Numeracy and Literacy: The Two Cultures and the Computer Revolution 92
14 · A Diary, Exhibiting Some Reasons Why There Is and Some Reasons Why There Should Not Be an Engineering Faculty Shortage 108

At Home

15 · Time Piece 139
16 · Dust Jacket Dilemmas 149
17 · A Little Learning 157
18 · Bullish on Baseball Cards 167
19 · Outlets for Everyone 170
20 · Is There a Big Brother? 182

At Play

21 · How to Balance a Budget 193
22 · Washington Entropy: Losses from the Energy Bill 197
23 · MX Decoded 206
24 · Letters to Santa 209
25 · Politic Prosody 214
26 · Modeling the Cat Falling 222

27 · These Goods Better Be Best 225
28 · Toys for Parents 231
29 · A Higher Options Exchange 234
30 · The Randys: An Immodest Proposal 238
31 · Metric Sports 243

Notes and References 249

Preface

At heart this book is about technology and technologists, but it avoids graphs, equations, and jargon. It begins with some of my experiences at Argonne National Laboratory, where I was an engineer during the energy crisis of the 1970s, and continues with observations of engineering students and engineering education today, from my present perspective as a member of the faculty of the School of Engineering at Duke University. The book continues with essays on some familiar artifacts and on some of the effects of technology on family life, and it concludes with a section of whimsical pieces that have a technological bent.

Most of the contents of this book first appeared, often in somewhat different form, in newspapers and magazines over the past decade. Places of first publication are noted at the end of the book, and I am grateful for permission to reprint these pieces now. Several editors supported my earliest attempts to write about technology and its effects from an engineer's perspective, and I should like to acknowledge especially Howard Goldberg of *The New York Times*' Op-Ed page, John Mattill of *Technology Review,* John Wilhelm, formerly of *Science 80,* and Steve Marcus, now of *High Technology.* I am also indebted to Tom Dunne of St. Martin's Press for making it possible for me to prepare a collection of my writing and to everyone at St. Martin's who has helped to make the book what it has become. It should go without saying, however, that I alone am to blame for any shortcomings or infelicities in this book.

Finally, I am grateful for the help and encouragement of my family. My wife, Catherine Petroski, has always been my most sympathetic and critical reader, and she has been no less so with this manuscript. My daughter, Karen, found the time during a busy summer to prepare illustrations. And my son, Stephen, as the reader will see, has provided me with much material.

—Henry Petroski
Durham, North Carolina
July 1985

Beyond Engineering

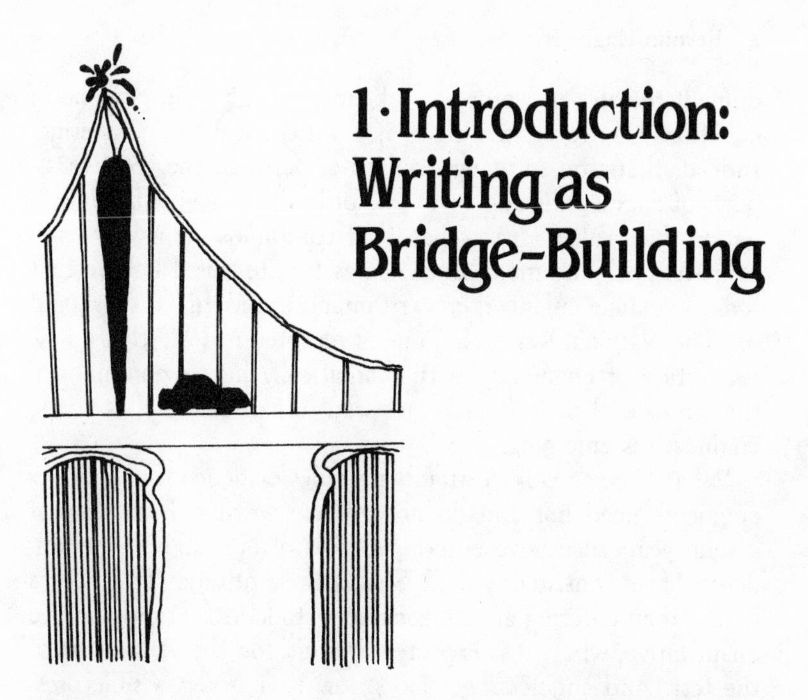

1 · Introduction: Writing as Bridge-Building

Engineering and writing are commonly thought to be mutually exclusive activities, and for a long time anyone who did both was considered a queer duck. This view has been so widespread that the vast majority of engineers seem to have accepted it as a given and have not expected themselves or their fellow engineers to be capable even of good technical writing, let alone any other kind. Samuel Florman, that rara avis who is an articulate apologist for the profession, once put it bluntly: "When engineers attempt to write creatively . . . the results are usually disastrous." Because of their predisposition to prejudge their own writing, few engineers have even tried to write outside their technical fields, and the expectation that engineering and writing do not mix has easily become a self-fulfilled prophesy.

For whatever reason, writing apparently does not come easily to a great many engineers, but then it comes easily to few writers, either. It is hard work, and problems in writing can be at least as difficult to solve in words as problems in engineering can be in equations. However, writing is not antithetical to engineering. Indeed, there are so many similarities between the creative aspects of engineering and writing—for both are acts of the imagination that realize ideas—that what confounds me is why engineers do not write more. But even as I write this, I believe that today's younger engineers *are* writing more, and this is as it should be. The National Research Council reported in 1985 that a new generation of engineers "with dramatically higher communication and social skills . . . as compared to past stereotypes of the engineer" is emerging.

What I see at Duke certainly provides evidence that younger engineers need not and do not eschew writing. For the past several years successive entering classes of our engineering students have consistently had Scholastic Aptitude Test scores higher than entering arts and sciences students, not only on the quantitative, where it is expected, but also on the verbal part of the test. And engineering majors now have a very visible presence in the student newspaper, where their letters to the editor appear frequently and for which they write columns regularly. Indeed, in recent years the most eagerly awaited and widely read weekly column was written pseudonymously by an engineering student. This and other successful journalistic endeavors certainly indicate that engineering students, at least on this campus, no longer exclude themselves from writing for a non-technical audience.

Yet established prejudices are as secure in the face of hard evidence to the contrary as are prevailing engineering and scientific paradigms. When one of my first pieces of non-technical prose appeared on the Op-Ed page of *The New York Times,* I was an engineer at Argonne National Laboratory, and the reaction of my colleagues ranged from curiosity to suspicion to incredulity. A former student of mine went so far as to ask in all seriousness if my wife, whom he knew to be a writer but *not* an engineer, did not in fact ghost-write the piece. I told him emphatically that she did not, but he seemed to remain incredulous.

The irony of this story is that my skeptical student himself did more than engineering. He enjoyed the non-technical pursuit of sketching and drawing, and he was proud of the portraits he had done and wanted me to see them. When I prodded him, he admitted that he had gotten good at drawing by drawing, drawing, drawing, and by looking, looking, looking at drawings, drawings, and more drawings. I told him that that was not unlike how I thought one learns to write—by writing a lot and by reading a lot of writing. I insisted that I preached only what I practiced, but his skepticism seemed to be total.

Another engineering student recently complained to the faculty at a senior gripe session that she felt two inadequacies in the four-year curriculum she was about to complete: She did not feel she had mastered the computer and she did not believe she had been taught to write well enough. While it may be that she was not inundated with four years of computing or writing assignments per se, she certainly had ample opportunity to compute and to write. She, like many students, wanted to be *taught*, whereas computing and writing, like all skills, are really *learned*—by doing. And those who do, often want to do more and better.

This particular student is a perfect example of the phenomenon that those who write best inevitably want to write better. She had been editor of the student engineering magazine and had just won a national student essay contest sponsored by the American Society of Civil Engineers. She clearly saw writing as an integral or concomitant part of her engineering activity, but she had not yet grasped that she would learn to be a still better writer principally by continuing to write, whether instructed or not, just as she would continue to grow as an engineer *only* by doing engineering, whether through textbook exercises or by practical experience. I suspect that this student was also a lot better with the computer than she admitted, perhaps even to herself, and that she complained about writing and computers because she wanted to do more with words both on paper and in the computer's memory than she felt ready to do at graduation. But that is so much better than thinking she knew it all, or that her need to learn had come to an end.

It is incontrovertible that one grows as a professional by doing. Indeed, we speak of *practicing* engineering, law, or medicine, and we believe there is no substitute for experience. By doing, we teach ourselves. By doing we expect to do better, and we do. We readily recognize that to excel at playing a violin or tennis requires constant warming up and practice. If we are taught anything about these things, it is good practice habits and the regimen of exercise. And the teacher does not so much teach as criticize, encouraging but always indicating how we could improve. Only in writing do so many expect somehow to excel without coming up through the minor leagues, and we seldom speak of the *practice* of writing. While one *practices* engineering, one simply *writes* (or hopes to) or does not.

Engineers as writers and engineering writing have long been written off everywhere. Even my dictionary clearly distinguishes technical from other writing in its definition of belles lettres: "literature as one of the fine arts; fiction, poetry, drama, etc. as distinguished from technical and scientific writings." Some engineers, however, especially the young and verbally more articulate, want to write even a technical article as well as they imagine an accomplished essayist creates an accomplished essay. They want to create a genre of *technical* belles lettres. (No matter what my dictionary says, I do not believe that is an oxymoron.) They worry about the grammatical function of equations, about whether the equals sign is a verb and, if so, whether singular or plural, and they speak of elements of style as seriously as do Strunk and White. While some may never wish to write for anyone except their fellow engineers, those of the new breed of engineer-writers do want to write well. What is important is that they value good writing, no matter what form it may take. And they do not wish merely to write good technical reports and articles; they want to be able to write well, period.

But old prejudices die hard, and a recent article in the magazine *Graduating Engineer,* ignoring the growing skills of its audience while giving advice to its young engineer-readers, went so far as to say: "Such traditional writing virtues as lively language, trenchant choice of word, and varied sentence structure count for little in technical writing. . . ." With attitudes like this rampant, it is no wonder that graceful style has remained the exception in technical writing. And without style, writing is without character, and it may as well be anonymous. We have been conditioned to expect and accept in professional journals the task of tedious reading, replete with passive voice and jargon galore. So if engineering students want models of style, they must read and write —to use the latest jargon of teachers of writing—not only across but also beyond the engineering curriculum.

Whether they like it or not, the activity of all engineers, not only academic ones, necessarily involves writing—both highly technical writing for their peers and less technical writing for their superiors. This was quantified by a recent study of a corporate research and development group. The study revealed that as much as one-third of an engineer's time is spent in writing and writing-related activities. While "writing" is seen to be an activity distinct from "engineering," writing and engineering can be viewed as two sides of the same practical coin. Engineering is, in its purest form, the creation of something new, and to be an engineer is to be, above all else, self-critical of one's creations. Sometimes the something created is a concrete bridge, sometimes it is an abstract theory of why concrete bridges work. But whatever the engineer does, he must anticipate how the thing he is constructing, whether in concrete or abstract terms, can fail to meet its objectives. This is little different from writing an essay or story, for the writer must anticipate the questions his reader will ask of the work, and to accomplish its end is to answer beforehand those questions. Just as the engineer must demonstrate to himself and to his colleagues the soundness of a bridge, so must the writer demonstrate to himself and his reader-critics the soundness of his metaphorical bridge.

John Gardner wrote of some of the difficult problems he faced in writing his novel, *Mickelsson's Ghosts:*

> If the novel is to begin where the story begins, then by the end of Chapter One Mickelsson must at least have located the house he will buy. We must know why he is hunting a house and what the hunt means to him—must understand why he hates living in town near other professors; must know, *by the firm proof of dramatized scenes.* . . . [italics added]

In other words, if it is to be credible, the thesis of the novel must be *proven* as surely as any mathematical theorem or engineering design. The writer must know what questions the reader will have about a character or a series of events or a place, and he must answer them before the reader even can begin to ask, just as the engineer must ask the questions the client or Mother Nature will pose of his construction.

When a bridge is needed to span a river, engineers and engineering firms can be invited to submit proposed designs, and the number of solutions to the problem of getting from point A on one bank to point B on the other side of the river can be as many as there are entrants in the competition. Conceptual designs for arch, truss, cantilever, cable-stayed, and suspension bridges made of varying proportions of concrete and steel can all result because different designers mix, in different proportions, their responses to the external and self-imposed constraints of function, constructability, economics, and aesthetics. The final choice may be based solely on the lowest bid, or, since the differences in cost can be of little significance, it may be the choice of a committee whose preference for one design above all others can be as subjective as is each designer's choice of what to submit in the first place.

Much of this effort could be saved, of course, by standardization—but how boring it would be if all bridges were of exactly the same design. Not only would the creative elements be absent from engineering, but also the aesthetic excitement of the built environment would be absent. Traveling would no longer be filled with the pleasures and surprises of finding different bridges in different places around different corners. It would be like reading the same plot in novel after novel or walking through a museum in which a single painting was reproduced in gallery after gallery, perhaps varying only with the size or the style of frame that it filled. The creative bridge designer, like the creative artist, has a portfolio showing the range of styles and media in which he works.

The great bridge engineer David Steinman prepared just such a portfolio in the form of a booklet published in the late 1940s. While the inside of the booklet documented many of the important projects to Steinman's credit, the cover carried a sketch of a proposed great suspension bridge across the Narrows between Brooklyn and Staten Island, "the last remaining gap in the regional traffic plan for New York City." The booklet explained the cover sketch:

> To close that gap with a monumental suspension bridge has been the goal of Dr. Steinman's career. He has been working on the development and design of this project for more than twenty years. It is his dream bridge.
>
> When constructed, the clear span from shore to shore will be over 4620 feet—a thousand feet longer than the George Washington span. . . .
>
> This, Dr. Steinman's dream "Liberty Bridge," will be the world's highest achievement in engineering. Spanning the gateway to Amer-

> ica, it will be a symbol of our free, vital civilization, a portal of hope and courage—an inspiring symbol of the spirit of America.

If Steinman seemed to see the bridge across the Narrows as the engineering equivalent of the Great American Novel, he was to be disappointed that his manuscript found no publisher. The gap in New York's traffic plan was to be closed, in 1964, but not by Steinman's dream. While his great Mackinac Bridge did connect the upper and lower peninsulas of Michigan, it was the concept of Steinman's rival, Othmar Ammann, designer of the George Washington Bridge, that was to become the Verrazano Narrows Bridge, with a main span exceeding that of the Mackinac by 460 feet. Ammann's bridge that graces the entrance to New York Harbor today is as different from Steinman's concept as a Norman Mailer novel is from one of rival Gore Vidal's. Though the main suspended spans of both Steinman's and Ammann's designs differed in length by less than ten percent, the bridges were no more aesthetically or structurally interchangeable than would be Mailer and Vidal novels of equal length on the same theme.

Fritz Leonhardt, the German bridge engineer, has written a lavishly illustrated book that is essentially a bilingual catalog of bridges around the world, with a welcome index linking many bridges to their individual designers the way novels are always linked to their authors. Leonhardt devotes an entire chapter of *Brücken/Bridges* to the question, "How is a bridge designed?" After listing the data needed at the beginning of the design project, including traffic requirements and topographical and geological details of the site, he describes the creative process in designing large bridges:

> The data . . . must be fully assimilated and remembered. *The bridge must then take its initial shape in the imagination of the designer.* For this process to take place, the designer should have first consciously seen and studied many bridges in the course of a long learning

> process. He should know . . . when a beam bridge, an arch or a suspension bridge will be suitable. . . . [italics added]

So the creative engineer imagines his design within the tradition of bridges he has "read" structurally as a novelist might imagine his design in relation to the tradition of the literature he has read. After several sketches of the designer's concept have, over a period of time, been drawn and criticized with regard to appropriateness for the site and function, the bridge engineer, Leonhardt says, behaves like an artist with his preliminary studies or drafts:

> The designer should now shut himself away with these first results, meditate over them, thoroughly think over his concept and concentrate on it with closed eyes. Has every requirement been met, will it be well-built, would not this or that be better looking or better for later detailing?

Only after an engineer settles on the basic design of his bridge, including sizes of major cables, beams, or girders chosen by experience and the roughest of calculations, does he even begin to think about detailed calculations of strength and resistance to such disturbances as wind and earthquakes. He then proceeds to "design by analysis," to flesh out his concept with more and more refined specifications for everything from major components to connecting details. No structure can be judged adequate without being realized, at least on the drafting board. And it is during this process of realizing the final design that the engineer discovers more and more about the behavior and inner workings of his bridge. The qualitative idea of spanning a river with such and such a design becomes the quantitative adventure of putting together the pieces to make an integral whole that will not collapse under the first traffic it carries or in the first storm it must weather.

Bridge-building is not so different from writing, and vice versa. The writer is the designer of a bridge of words. Just as each engineer will come up with a different bridge between the same two points, so will each writer produce a different poem or essay or novel on the same "subject." And the writer no less than the bridge engineer will consider the form, tone, voice, and even time of composition that he sees fit to the task. The essayist who writes every essay to a formula would be no more considered an artist than the engineer who made every bridge an enlargement or reduction of his last.

In *The Philosophy of Composition,* Edgar Allan Poe reflected upon the process by which creative writing is done, and he sounds not unlike Leonhardt on designing a bridge:

> Nothing is more clear than that every plot worth the name, must be elaborated to its denouement before anything be attempted with the pen. . . .
>
> I prefer commencing with the consideration of an *effect.* Keeping originality *always* in view—for he is false to himself who ventures to dispense with so obvious and so easily attainable a source of interest —I say to myself, in the first place: "Of the innumerable effects, or impressions, of which the heart, the intellect, or (more generally) the soul is susceptible, what one shall I, on the present occasion, select?" Having chosen a novel, first, and secondly a vivid effect, I consider whether it can be best wrought by incident or tone—afterwards looking about me (or rather within) for such combinations of event, or tone, as shall best aid me in the construction of the effect.

How is this different from Leonhardt's need to riffle through his mind's catalog of bridges that accomplish the desired goal? Writing no less than engineering consists of ideas to be realized.

Robert Frost said writing free verse is like playing tennis without a net. The artistic challenge of both literature and engineering is to use the net to one's advantage, to push against the envelope of form. The requirements that a bridge be so many feet long, so many lanes wide, and meet so many volumes of federal regulations are not that much different than the constraints that a sonnet have fourteen lines, five iambic feet per line, and a predetermined rhyme scheme. Within such rigid rules and regulations, William Shakespeare constructed sonnets as different from each other as the Brooklyn Bridge is from the Eads, or as Steinman's would have been from what Ammann's Verrazano Narrows Bridge is.

Writing is also like engineering in that the very doing of it is an act of exploration and discovery. To analyze a bridge's behavior as a total structure is often to find things that were not anticipated in the imagined design. These things might be its propensity to sag too much under traffic or to misbehave in high winds, but they might also be ideas that bring out the essence of bridgeness more clearly than had been done before. Such are the fruits of research and development, exploration and discovery.

Discovery in writing can be no less striking. There is not one of my technical reports or papers in which I cannot remember coming to a greater understanding during the writing itself of the subject about which I set out to write. The very act of "writing up results" unifies those results of calculations and experiments for me in ways that thinking about them in abstract or outline form does not. The test of an engineer's understanding of his own theoretical work often comes when he undertakes to commit it to words in a piece of technical writing. The first draft can serve the purpose of making clear what is *not yet* known clearly enough. Subsequent drafts reveal what has been learned in the act of writing. Writing in general, I have found, is no less an act of discovery than I have experienced in writing technically. Flannery O'Connor once said that she didn't know what she thought until she saw what she wrote.

To be an engineer is to have to look at details. My own engineering work often involves focusing my analytical and literal attention on what happens within a thousandth of an inch of the tip of a single crack that might develop in the steel in a bridge or the concrete in a dam spanning a chasm a mile wide. The world is full of figurative as well as literal cracks, bridges, and dams—and one can explore them and their details on a typewriter as well as on a computer. In doing so I have discovered things and ideas I am not sure I would ever have known, for the determination to finish a piece of writing has driven me to persist to find out about things I might otherwise have let lie. I did not really know how large giant windmills were projected to be until I needed that information to argue in an essay why the failing prototypes could be expected to fail. I did not know the exact thickness of the dust jackets on the thousands of books I had handled until it became natural in another essay to ask how much shelf space a library might save by removing the jackets. I did not know exactly what I thought of electronic toys until I agreed to write about their impact on our family life.

I write about a spectrum of subjects beyond engineering because, besides being an engineer, I am also husband, father, neighbor, citizen, pocket philosopher, and closet clown. My stage is the page, and on it I enjoy writing as much as my colleagues enjoy performing in drama clubs or on the tennis courts. Words are my leisure, yet I am no more content only to read than are the athletically inclined content only to watch spectator sports. Why I choose to write while other engineers play golf or collect stamps is no more clear to me than it is to them. And I write in a variety of forms and attitudes for the same reasons that the tennis buff tries a new racket and alters his serve or his swing. We are fascinated with the objects of our activity, and we adapt to the conditions of the court and the expectations of the spectators.

What is in this book are the fruits of about ten years of extracurricular writing. Almost all of the pieces were written after my wife and children had retired and the house was quiet. I have been absorbed in working on essays at my desk as I imagine my woodworking colleagues have been in working on cabinets in their basements. And like them I have worked up to the ambitious pieces after spending a long time practicing on the simpler. As the woodworker begins with lamps, shelves, and parqueted trays, perhaps, so I began with letters, parodies, and poems, writing hundreds of mosaic verses in the late evening when the meter and rhyme and words filled my head and involved my senses.

I have never been much for prophesies of the kind that tell engineers they cannot write, and I have never seen any reason for engineers to settle for any less proficiency in manipulating words and sentences than they are expected to have in manipulating mathematical symbols and formulas or the specifications for an engineering structure. My earliest attempts at writing may have been as shaky as a sophomore's bridge, but I enjoyed the process of writing and imagined I was always improving. When I eventually received positive feedback from editors, I continued to write though free time became always more dear. I hope that my incredulous student has likewise continued to draw, for I believe that the mental discipline he no doubt gains from drawing will make him a better engineer. I believe creative activities like writing, drawing, and engineering are but different manifestations of our single humanity, and do not necessarily come from irreconcilably opposed sides of the human brain.

Just as the call for bids for a bridge from A to B gives rise to a variety of structural solutions, so can the intention to write from a thesis to a conclusion an essay or a whimsical antidote to the evening news give rise to flights of imagination and a variety of bridges of thought. I hope, however, to borrow Marianne Moore's idea of poetry, that these are imaginary bridges with real roads on them. And I hope the journey through this book will be as pleasurable as a drive over the wonderfully diverse bridges of the world.

But as each bridge is a portrait of its engineer, so this book is at heart me. Whatever the structure of a book, or even an essay or technical article, it ultimately is a manifestation of the author's experience, perspective, and imagination. In technical writing the first person singular is frequently eschewed, lest it be pluralized by co-authors, referees, and editors. This only aids the illusion of objectivity. There can be no such illusion about the pieces in a book such as this, however. Montaigne admitted up front and candidly to the readers of his *Essays* that he himself was the matter of his book, and I am the matter of mine. I hope my immodesty might help dispel the myth that engineering and writing are immiscible.

At the Laboratory

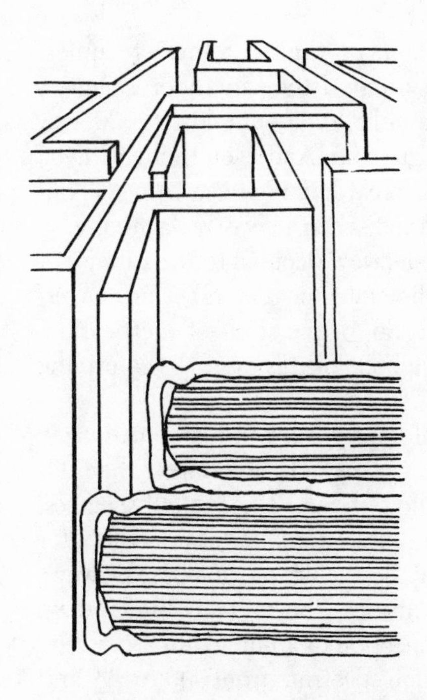

2·Of Two Libraries

I am a regular patron of two libraries: a highly technical one at the research-and-development laboratory where I am a member of the staff, and the public library of the village in which I reside. For years I have browsed among the shelves of these libraries, often visiting the first during lunchtime and the second after dinner that same evening, and I have been struck repeatedly by the minuscule overlap in their collections.

This fact, along with some other characteristic differences between the two libraries, is a constant reminder to me not only of the divergent objectives and concerns of the technical and non-technical communities but also of the isolation of specialists. The libraries further symbolize to me the obstacles that must be overcome for communication to exist among scientist, engineer, and citizen—or even among scientists and engineers themselves—and the necessity of cooperation between specialist and generalist.

As if to emphasize their differences, the laboratory's technical library uses the alphanumeric system of the Library of Congress to classify its books, while the public library employs the Dewey Decimal System, with which I grew up. And even though I lived through the trauma of using the great library at the University of Texas at Austin during the period when it was reclassifying its three million volumes from the Dewey Decimal to the Library of Congress system, I still have a difficult time remembering conversions like QA612.19 = 514.22. So, in the spirit of metrication, when it is necessary to refer to a book in this essay, I will use the decimal system.

The collection of a technical library is by and large cataloged under only two of the ten major classes of the Dewey: the 500s and the 600s, which contain the pure and the applied sciences, respectively, though the rapid proliferation of books in computer science is beginning to overflow the shelves of the 001s, Dewey's slot for "knowledge and its extensions," beyond any limits imagined even a decade ago. The collection of a public library, on the other hand, spans the classification spectrum from 000 to 999 and goes beyond it by throwing in a massive amount of creative and imaginative writing, usually under the numberless designation *F*, for fiction, and a wealth of human experience, frequently under *B*, for biography.

Technical libraries can be imagined to be stressing detail by amassing information to more and more Dewey decimal places. This is literally as well as figuratively true, and call numbers like 620.1920426—this one actually assigned by a technical library to an English translation of a Russian monograph published in the Netherlands under the title *Fracture Micromechanics of Polymer Materials*—are not uncommon nowadays.

It is rare that I come across such refinements in call numbers at my public library, however. That library is more interested in providing broader, order-of-magnitude approximations to the needs of its patrons. It also provides, through biography, fiction, and its poetic relatives, a view of the world and its people that is qualitative rather than quantitative, subjective rather than objective. The public and technical libraries are the forest and the trees of book collections. Thus the technical and secular goals seem clearly to be divergent, though they need not be incompatible.

In order to understand the exotic trees in the delicate ecosystems of their specialized sub-forests, the laboratory's scientists and engineers constantly demand more and more specialized literature: esoteric monographs and obscure journals with press runs in the hundreds and proceedings of international meetings announced to, attended by, contributed to, and published for no more than fifty or a hundred specialists. Its users are so specialized that the laboratory's library (officially known as the Division of Technical Information Services) actually comprises ten branch libraries for a population of fewer than two thousand scientists and engineers.

Each of the laboratory's branch libraries concentrates on a narrow discipline, like chemical engineering (Dewey's 660s) or high-energy physics (principally the 530s). The branches are usually located in the same buildings as their users, and thus there is little need for, say, a chemical engineer to leave the Chemical Engineering Building, except to go to lunch at the cafeteria—where he usually sits at a table of chemical engineers. Such an isolation of the specialist was probably not the intention of the laboratory's system of separate buildings and branch libraries, but it is one of its effects.

Because library space does not keep pace with acquisitions, the laboratory's branch libraries would have little room for the vast numbers of new books they acquire if the librarians did not encourage the scientists and engineers to keep books most relevant to their work in their offices. Thus, twig libraries sprout between the desks of staff members sharing offices. The books are never overdue, and researchers seldom receive recall notices for the volumes on their private shelves, because few if any other researchers want to read those same details. While the difference in the call numbers on the two sides of a bookshelf in an office may only be measured in the decimal places, the perspectives represented there may be as divergent as the views through the two ends of the same telescope.

If one tries to rise above it all and look through the roofs of the laboratory buildings that house the separate branches of the library, one sees several sets of separate labyrinths of book stacks. From this vantage point one can see physicists and chemists, mathematicians and engineers executing mazes of shelving each to track down an obscure reference to an obscure reference. The paths of the researchers seldom cross. If they are not in separate buildings, they are in separate aisles, looking at the same physical world between the blinders of different monographs.

Even in a public library one's vision is channeled between the stacks, and it can be difficult to see the library for the books. However, if one removes some books from the nearest shelf, he can see through to the next aisle, at least, and if he were (metaphorically) to stack books horizontally to make steps upon which he may ascend, he might begin to gain perspective on the many different ways of looking at the same world. Perhaps this is what Newton imagined himself doing when he spoke of seeing further than his contemporaries by standing on the shoulders of the giants who preceded him.

A pile of monographs with call numbers identical out to the fifth Dewey decimal place does not reach very high, however, and the specialist who stands upon it can barely see over the topmost shelf of the special aisle of the special library that he frequents. From this perspective he cannot look down even into other parallel aisles, and his line of sight is limited by the outside walls of the narrow library. Only his fellow specialists, those browsing nearby, will see him, and they all will know that he is standing not on the shoulders of giants but on the horizontal spines of monographs, for they have all tried the same trick themselves. So to his colleagues two aisles away he may as well be out to lunch at his building's table at the cafeteria.

Users of narrow branch libraries sometimes find themselves out on a limb, however, and their only safe way back to earth is by way of the trunk, or central library, which contains the general scientific and technical reference works and the books on policy and law whose leaves cast shadows on all the books in all the branch libraries.

The central library houses the office of the head reference librarian, to whom the scientists and engineers go with their most obscure and vexing bibliographical problems, problems too vague or too general for the branch librarians. The head reference librarian stands rather than sits at the desk in his office, for he is (literally) a midget in a room of oversized furniture. Yet he has the perspective of a giant and the memory of an elephant. He puzzles over fragments of references the way an archaeologist contemplates fragments of bone. And as surely as the archaeologist will infer a dinosaur from a femur, so this lilliputian librarian will nail down a book with a fragment of a page—though it may take time.

He dogs the unfound book the way a mathematician does the proof of a theorem, a physicist the track of an unseen particle, an engineer the optimal design. And with the help of the stooping researcher he is trying to help, the small librarian is able to riffle through this book and that from the uppermost shelves that he himself could never reach, and thus he can function in spite of what might have been a handicap. Indeed, the midget has led giants to realize the implications of their own discoveries.

Most citizens, who are neither midgets nor giants, do not have need for a research library, let alone a branch or a twig research library. They use their local public libraries, where fact and fiction, science and poetry, are never shelved very far apart, and always in the same building. And no matter how small the building, it is always more a microcosm than a microscope slide, for Dewey's continuum is housed under one roof. The common man can move freely from art to philosophy to sociology to poetry to history to science to technology. There are few monographs to stand upon, or to separate one aspect of life too far from another.

When the time comes for public discussion on broad, complex issues like energy policy, the differences between public and research libraries exacerbate misunderstandings and complicate debate. Whereas the specialists, who after all will be called upon to effect any technological fixes, are deeply read in the narrow technical aspects of the problem, the intelligent public—which may include the off-duty specialist himself—is often familiar with the broad, social aspects of the issue. How to reconcile the different points of view, how to reference the two libraries, is one of the most difficult tasks of our time.

Newton is reported to have said near the end of his life that he seemed to himself to have been only like a boy, playing on the seashore and diverting himself now and then in finding a smoother pebble or a prettier shell than the ordinary, while the great ocean of truth lay all undiscovered before him. The smoother pebble or prettier shell might also have been a more elegant monograph or a finer book of poems, and the ocean a library. Which of us has not been elated at finding a gem of a book in the library; and who has not been humbled by the miles of volumes, shelved beyond the span of our lifetimes, with newly catalogued volumes forever hitting the new-book shelves, like the endless waves bringing new shells to the shore?

Yet Newton, whose own library contained more than twice as many non-scientific as scientific books, more books on classical Greek and Latin literature than on mathematics, more on what was then modern literature than on physics, more on travel than on astronomy, and more on theology than on alchemy, chemistry, mathematics, medicine, physics, and astronomy combined, still found time to write his *Principia*—a monograph and a new worldview at the same time—and oversee the reorganization of his country's mint.

At their best, specialization and generalization are not mutually exclusive but complementary qualities, just as the research and public libraries are complementary sources of information and inspiration. The books necessary to write a balanced history of the past, an accurate worldview of the present, or a successful proposal for the future are not to be found strictly in one or the other of the two libraries, for neither one alone has all the books, nor all the knowledge.

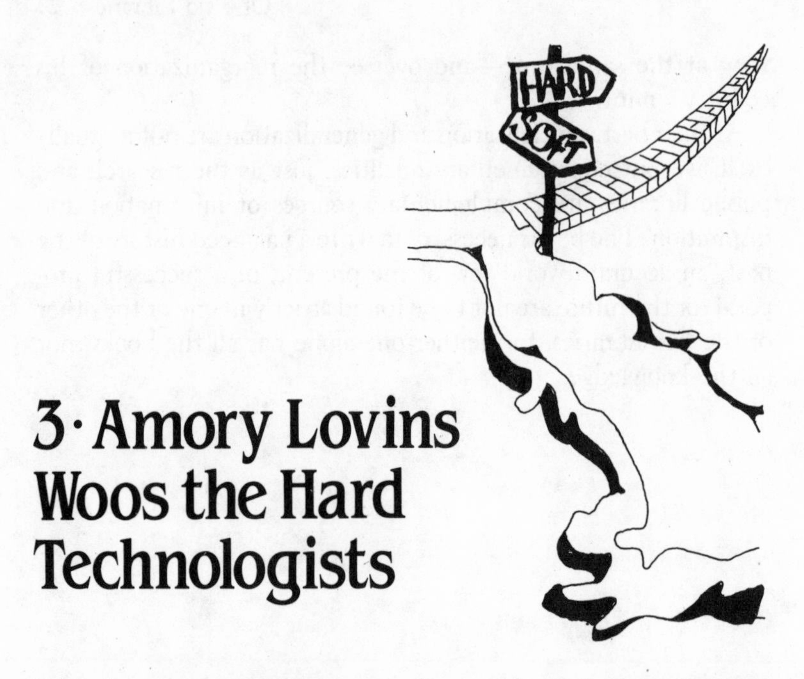

3 · Amory Lovins Woos the Hard Technologists

Like the energy paths he wants the world to take into the next century, his very name—Amory Lovins—is soft, soft as the whispers in this auditorium, where three hundred hard technologists are waiting to hear him speak this morning. As I relax in one of the cushioned theater seats, I think of Cleveland Amory saving Arctic animals, I recall my Latin *(amo, amas, amat),* and I remember the flower in the gun barrel on the steps of the Pentagon. Flower power and windmills are soft, modern armies and nuclear power are hard, we all would soon be reminded.

Having survived the 1960s, twenty-four-year-old Amory Bloch Lovins resigned in 1971 from his position of junior research fellow at Merton College, Oxford, to become British representative of the environmentalist group, Friends of the Earth. Since then he has become probably the most articulate spokesman for alternatives to the hard-technology approach to future global energy needs.

This morning, just eight years since his dropping-out from the hard academy to choose a softer path than the conventional ones through the groves of academe, Mr. Lovins is addressing these patient scientists and engineers—the vast majority of whom endured four, five, or more years of post-graduate regimen to earn doctorates in their respective fields—at this national laboratory that is distinguished for its role in advanced nuclear power research and development. Here, where budgets, which have yet to be stripped of their hard, sexist terminology, are quoted in man-years, the effort being expended on this seminar is, to a manager at least, nothing to be sneezed at.

Beginning at nine o'clock this morning, three hundred hard scientists and engineers excused themselves from their desks laden with calculations, data, and computer printouts to assemble in this largest auditorium in the laboratory. This is scores more than turned out to hear Edward Teller, arguably among the hardest of technologists, talk in this same room several years earlier. Interest in Lovins was correctly anticipated to be so great that admission is by reservation only, and tickets were distributed on a first-come, first-served basis. The event, with questions, would extend almost to noon, which amounts to about a nine-hundred-man-hour (almost half a man-year) total laboratory effort to give Lovins an opportunity to filibuster and proselytize.

While Lovins is about to put on a one-man show not very different from that of a self-assured author on a television talk show, the audience really deserves more credit than to be simply grouped together as half a man-year. These three hundred individual men and women really share only one common feature—their place of work. For here are engineers who have built energy-efficient houses in the woods, using heat pumps and solar collectors as best they know how. Here are physicists who eat, sleep, and breathe thermodynamics and who are forever looking for the theoretically best choice for a future energy source. Here are chemists who know how difficult it is to burn dirty coal cleanly.

And they all know about synfuels and wind and waves and all the other utopian energy sources. They work on them, too, but a good half of the audience does spend at least *some* of its time on nuclear power research and development, for it is that mission that pays the bills—with enough left over to pay handsome honoraria to distinguished speakers.

Lovins is finally introduced by an associate laboratory director who looks as if he could be an associate professor of English introducing a poet at a poetry reading. The show is about to begin.

This is the setting: There is, naturally, a stage in the auditorium, and at stage rear is a huge projection screen upon which is focused the beam of an overhead projector. The projector is sitting on a small table placed front-row center on the auditorium floor. Lovins is not on the stage but is sitting beside the table. He is sitting during his introduction and will remain so until he rises to make a point to a questioner after his talk. Lovins's face is only partly visible in the light of the projector, but then only to the long-necked in the audience. He will lecture *ex cathedra* for exactly one hour in the darkened auditorium. A blunt person would call his style rude.

Before beginning his talk proper, Lovins holds up in the projector's beam, which he uses theatrically, a copy of his book, *Soft Energy Paths.* This two-year-old book is the subject of the morning, and virtually everyone in the audience has read it.

Some of Lovins's transparencies are very professional, as good as any prepared by the graphic arts department of a Westinghouse or a Mobil to push corporate energy approaches. Like the representatives of those corporations, Lovins uses color cleverly in his visual aids: red for (hard) nuclear, green for his (soft) alternatives.

Most of the viewgraphs are boring, however. Too many familiar numbers and too many old examples from the book. There is nothing new, save an updated statistic here and there. The talk is longer than it has to be, but nobody leaves. Lovins did write a powerful and persuasive book; he must have more to say.

Three microphones for the audience are set up in the aisles. Those with questions for Lovins are instructed to assemble behind the closest mike and to await their turns. The first question comes from the middle aisle.

"I just finished building a house, and I tried to put as much thought into its energy efficiency as I could. Now I find I don't save all that much energy, and I can't see how to save any more unless I block off the double-glazed windows completely in winter. Do you think the energy-saving goals you show on your charts are really fair?"

"How much insulation do you have in your ceiling and walls?" asks Lovins.

"R-24."

"You don't have enough. You should have R-36. You live in an energy sieve."

"All the handbooks said R-20 was more than adequate."

"Your windows are only double-glazed. You live in a sieve."

"Double-glazed windows are supposed to be sufficient."

"Don't believe manufacturer's specifications. You built a sieve."

"What do you suggest I do?"

"Two mechanical engineers in Manitoba have just written a paper describing a house they built to withstand the Canadian winter up there. Here is a chart showing the specifications . . ."

"But they have less window area than building codes here allow."

". . . You live in a sieve."

For the next hour and a half or so there is less dialogue, as Lovins makes it clear that he is here to talk and not to listen, to poke holes in everyone's argument, but to admit none in his own. Not until the first and only female questioner, with a hint of belligerence in her voice, persists into lunchtime in her skepticism of Lovins's energy path, will there be another extended interchange of views. In between will come a succession of questions generally expressing sympathy for Lovins's argument, but admitting professional doubts that his goals are realistic.

The questioners by and large are engineers and scientists who have worked on complex and futuristic energy systems for most of their professional lives. They know the heat transfer properties of liquid sodium and of household tap water. They know the technological problems with solar energy and with windmills. They know, and they politely confess them to Amory Lovins.

These polite doubters are not sartorially distinguished, and most stand with poor posture, waiting their turns to speak. Unlike tweedy Lovins, who now is moving freely and effectively back and forth in front of the stage as if lecturing to a class on classical thermodynamics, the technologists are queued up behind stationary electronic instruments over which they will not demonstrate mastery when their turns come.

In *Soft Energy Paths* Lovins uses Robert Frost's poem "The Road Not Taken" to symbolize the more desirable soft energy paths chalked out into the grassy future. "The Road Not Taken" might also serve as the explanation of how many of these engineers and scientists came to be where they are this morning. Many simply followed a quiet academic road that, beyond the bend, led them willy-nilly into nuclear power research and development.

For most of these former science and engineering students, who are typically now in their mid-to-late thirties, as for many people generally, the first fork in the road was encountered in their choice of college major. If the student was a male in the late 1950s, liked math, and liked to work on cars and model planes, he was very likely advised that all roads would be paved with gold in four or five years if he chose engineering.

Professors in the early 1960s, with swelling research contracts and grants, enticed hordes of young bachelors to stay on the academic road in full-time graduate programs. Most students took the path of convenience and of least resistance and wrote theses on the same research problems the professors paid them to attack. And those problems were, of course, in the areas funded at the time: defense, aerospace, and nuclear. Sooner or later, the young men were doctors of science and philosophy with esoteric dissertations they would spend the next several years trying to publish and outgrow. Little did they know how difficult the latter could be, and little did they realize that the more degrees one has, the fewer things one is allowed to do. These Ph.D.s knew as much about philosophy as most philosophers do about engineering.

Yet the world is made as much of philosophy as of concrete and steel. Energy paths are philosophical constructs of the masters of business administration and of macroentrepreneurs. The reason a nuclear path, rather than one lined with windmills, has been laid out is the same reason that an interstate highway system, rather than a well-maintained railroad network, was the focus of legislative attention for several decades.

Now these humble engineers and their less humble colleagues, the scientists, stand at the microphones as if waiting at a religious crusade. Although Lovins's talk was uninspired, these hard technologists want to believe, and they offer themselves one after the other to the man in tweeds. He takes their predictable questions and runs with them through appendixes to his talk. To this question he produces this viewgraph; to that, another. Time and again he answers the questioners with the same hard technological jargon they themselves use at esoteric conferences to defend the arbitrary and belabor the obvious. He manipulates the data as a faith healer does the cancer of a true believer.

The infantry is looking for alternatives the generals have already rejected. Lovins seems to sense this, and he is really recruiting with his books and his talks. He knows that a lone poet cannot lead the world down the soft-technology path without an army of hard technologists paving the way. It takes the same engineering drudgery to develop a reliable windmill society as a safe nuclear one.

If the engineers and scientists are not as articulate as Lovins, they are at least as sincere. Many are already devoting their best professional years to soft energy paths by different names, and many others are sharing their time among nuclear fission and fusion and coal and synfuel and solar energy. Among them are the pioneers who are building energy-efficient homes on their drafting boards and on their subdivision lots. They are the guerrilla forces Lovins will need to secure a soft energy path through a future where hard technology already claims the right of way for its high-voltage transmission lines.

But if Lovins hopes to lead hard technologists along the less-traveled path, he will have to demonstrate more endearing qualities than he did today. He appears to be too hard-nosed about soft technology, and technologists, like others, suspect that people use the most words when they are the least sure of what they are saying. The audience diminishes during the prolonged question-and-answer session. This will not do, for one cannot hope to lead an energy revolution, hard or soft, without a benevolent corps of engineers.

4 · Soft Technology Is Hard

As with Claes Oldenburg's soft electric fans and other pop-art sculptures of the 1960s, there is an incongruity of scale and medium among the hard prototypes of soft technological windmills for the 1980s. Known as "wind energy conversion systems," or WECS in awkward acronym, they will evolve into wind turbines whose blades will span a football field and put windy little towns like Boone, North Carolina, where one of the world's largest windmills was to generate electricity for the local power grid, on the map. However, even the early prototypes pointed up the critical role engineers and other hard technologists have in charting viable paths, soft or hard, into an unmapped energy future.

In 1973, in the wake of the Arab oil embargo but before the soft-path metaphor entered the vocabulary of concerned citizens, the National Science Foundation sponsored a project to explore the potential of wind power. Whether the wind turns a blade on a stationary turbine or a propeller pulls an airplane through a relatively quiescent atmosphere is six of one and half a dozen of another to an engineer concerned with developing reliable and efficient blade designs, and NASA's Lewis Research Center was a logical choice for the project's early home. However, the Department of Energy subsequently was formed and took over the effort to convert free wind energy to electricity on a practical and economical scale that would ultimately be attractive to utilities. Hence the dual logos of NASA and DOE appeared on the Boone wind turbine, which sat on a 140-foot-high tower atop the Blue Ridge Mountains and was designed to generate two thousand kilowatts of electricity in twenty-five-mile-per-hour winds. To some, it was a beautiful sight; to others, the noise was so aggravating that the hours of operation were curtailed and eventually the project was abandoned and the equipment sold at auction.

The so-called soft energy paths are of course those along which society relies upon elemental sources, such as water, wood, and wind, to generate power, heat, and electricity, preferably on a decentralized basis. This is in contrast to the hard paths, along which energy is produced primarily from the likes of coal, oil, and uranium at large centralized power-generating stations. The soft and hard paths have competed for the energy traffic of the future, and, since they are largely parallel roads, the principal attraction of one over the other appears to be the tariff at the toll booth.

At present the hard path, generally believed to be open and paved all the way through the next century, is not only the quicker but also the more reliable and cheaper route for most. Still, some who prefer the quaintness and pace of old-time rural America voluntarily choose the soft path, much as some might prefer to take U.S. 40 instead of Interstate 80 between Chicago and New York. However, those same travelers would not like to hear the drone nor see the towers of wind turbines every few hundred feet along the Blue Ridge Parkway.

Thus, advocates of hard energy paths are not just tilting at windmills when they accuse soft-path proponents of being a little quixotic. To meet modern power requirements economically and conveniently, alternative energy systems must be conceived on a large scale. It would not be practical or economical, nor would it be pleasing to the eye, to put a small windmill, like a television antenna, on every rooftop. Tomorrow's windmill designs must be real, and not just imaginary, giants that tower over hundreds of families and deliver power for under a nickel a kilowatt-hour.

While the soft energy path may be lined with tulips and quaint Dutch windmills for some, for others it passes right through the giant steel structures supporting WECS, much the way paved roads tunnel right through giant redwoods in Sequoia National Park. Yet the tips of the blades of this new breed of windmill will reach higher than any sequoia, and when such blades rotate at even the relatively slow rate of twenty revolutions per minute, the speed of a blade tip approaches two hundred miles per hour.

One can imagine the potential consequences of such a blade breaking over a community of homes it is intended to serve. The windmill would be tilting for real at its defenders, and the crash of the errant blade would not be unlike an Indianapolis 500 car going out of control into the grandstand. Yet the severe mechanical stresses on a massive blade in ever-changing weather are precisely the conditions under which cracks develop and grow in the finest of steel and, sometimes, catastrophically sever blade from hub. And even should they not break away, such large windmill blades can conceivably pose threats to migrating birds and other flying creatures, interfere with aircraft and radar, prove unsightly in excess, create unacceptable noise—as in Boone—or interfere with television reception, as they have on Block Island, where another large windmill has operated. Thus, the implementation of a soft energy path can affect the environment in ways as complex as those of any coal or nuclear plant, though the individual consequences may not be as great.

Not unlike Oldenburg, who took the common electric fan and magnified its vulnerability in soft vinyl over foam rubber, engineers have been taking steel and concrete and molding the soft concept of wind power into a technology as hard as wide-bodied aircraft and nuclear reactors, and as vulnerable to design flaws. This was to be expected, for a nature trail in our industrialized society can go only so far before it has to cross the hard realities of economics and accountability and development. Perfection comes easy in soft dreams; it comes hard in engineering realities.

As with our vehicular roads and highways, the soft energy paths will be laid out and maintained eventually under the supervision of engineers, who will learn from today's prototype windmills how to design aerodynamically optimal and structurally sound blades that will withstand not only decades of use but also the vicissitudes of weather and time. The engineers will build large windmills not for their own sake but for the sake of the economy and society, and the engineers will be both damned and praised for their hard sculptures in concrete and steel, with blades of whatever proves to be the most durable and reliable choice.

5 · The Gleaming Silver Bird and the Rusty Iron Horse

Apart from World's Fairs and interstellar fantasies at the movie theater, air travel is as close to a futurama as most of us get. Not only are jet airplanes the nearest thing to space travel, but also the appurtenances of flying have all the ingredients for tomorrow's most popular disco spot. Modern airports are distinguished by their clean, dynamic architecture, fashionable people, and technological gadgets galore. Their vast expanses of chrome, glass, and plastic reflect all this and seem to concentrate it on the exhilarated individual bounding to his flight.

Railroad stations, on the other hand, are relics that bring to mind crowded courthouses and empty museums. They are cavernous and somber structures full of anachronistic advertising and derelicts living in another time. The grand urban terminals of yesterday, if they are not already demolished, are now depressing places of marble lost to grime where one sees fashion mainly in the magazines in which he buries his nose while waiting for his train to be announced—if the public address system is working properly.

Such contrasts, such extra-technological differences that have nothing to do with schedules or safety or energy, seem to me to be at the heart of the emergence of air travel at the expense of the railroads in this country. The airplane has been for some time now the available symbol of the future, the precursor of space flight. The railroad train, linked in history with the (ho-hum) conquered West, has always been chained to the past; it has been the iron horse forever tethered to the steel rails anchored in the ground. The railroad keeps us down-to-earth and constantly reminds us of the not-always-pleasant interrelationships between technology and society.

The airplane lets us fly and forget. We are as gods, even in coach class, attended by young, smiling stewards and stewardesses who bring us food, drink, and entertainment. From the window of the airplane we marvel at the cities far beneath us, at the great land formations and waterways, and at the clouds. Political boundaries are forgotten, and the world is one. Everything is possible.

Nothing is forgotten on the train, however. The right of way is strewn with the detritus of technology, and technology's disruptiveness is everywhere apparent. Outside the once-clean picture window of the train, which has probably slowed down to pass over a deteriorating roadbed under repair, one sees not heaven in the clouds but the graveyards of people and machines. One cannot help but notice how technology has changed the land and the lives of those who live beside the rails. The factory abandoned is a blight not easily removed; the neglected homes of myriad factory (and railroad?) workers are not easily restored.

To ride the train along the Northeast Corridor or through any old industrial section of our country is to see the human costs of technological growth. The main line affects the lives of its neighbors. Row upon row of wood-frame houses back up to the tracks and rattle on schedule with the passing trains. Children eager to get to the other side of the tracks wave the trains on. School buses and cars and trucks and even pedestrians wait at grade crossings; track crews wait to return to the rotting ties. The train disrupts them all.

Unlike airplanes, which fly above it all and whose only tracks are their ephemeral contrails and the blips on radar screens, railroad trains cannot change altitude to escape turbulence. The engineer cannot come on the public address system like the airplane captain and blame the rough ride on the weather. When the rails are rough the passengers feel it, and they are reminded of the steel on steel of the rails and wheels at every joint in the tracks. There is an inescapable relationship, however uncomfortable, between man and machine that is absent in the airplane. The *clackety-clack* of wheel on rail is like the *tick-tock* of a grandfather clock, reminding us that time is passing, that we are growing old. The whir and whine of the jet engine gives, like the hum of the electric clock, an illusion of timelessness.

The railroad in this country is an endangered species not because it is technologically obsolete but because it is technologically blunt. It does not misrepresent or mask the social and emotional costs of technology. Railroads keep us in the present and the past and show us things as they are and have been, not as they might be in a futurama.

To journey on a train through a major industrial region of this country is to ride through exhibit after exhibit on the history of technology. Mills, foundries, plants, factories—they are all along the main line that brought them raw materials and distributed their manufactures, before the interstate highway system made it possible to locate anything and everything elsewhere and everywhere. Too frequently now beside the railroad the earthy materials of which modern technology was constructed—brick, stone, wood, steel—lie in piles of debris and remain only as skeletons of structures that once pulsed with energy.

Amtrak once projected that it would cost several billion dollars to complete improvements along the Northeast Corridor between Boston and Washington, but welded rails, concrete ties, and other technological fixes for smoother, faster, and safer rides would not change the scenery or the reality along the tracks nor the conditions in the waiting rooms of large urban railroad terminals.

Should Amtrak survive, rail travel along the Northeast Corridor might once again be safe and energy-efficient. Perhaps it might even herald a new era in which we do not shut our eyes to the sociological costs of technology. For it is the railroad train, and not the airplane, that ties us to our technological past, and it is along the railroad tracks that we can view the historama of technology past that gives us lessons for the future. If we choose not to look out the window of the railroad train now, we may be condemned to repeat our technological mistakes and find our airplane travel less glamorous in the future. But who can imagine that now?

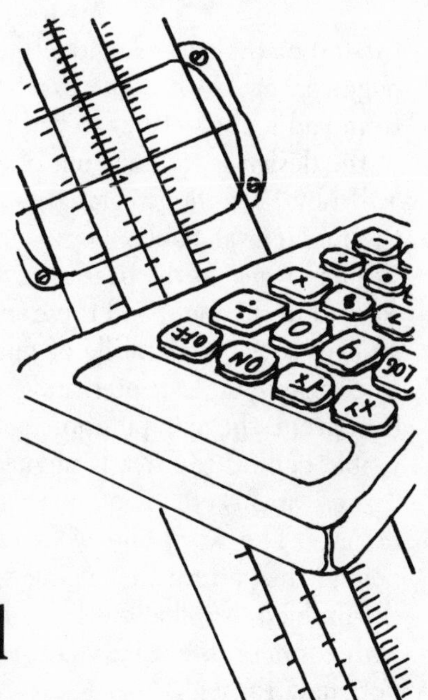

6 · Reflections on a New Engineer's Pad

Computation pads of pale green quadrille paper are to engineers what long yellow legal pads are to lawyers and what small white prescription pads are to doctors. There is a familiarity of size, color, and texture that the practitioner comes to feel comfortable and correct with, and writing professionally on anything else can be distracting. Today a new design on the top page of the fresh "Engineer's Pad" on which I am working has distracted me from setting down the equations and calculations pertinent to the behavior of the cracked structures—the large pipes in a power-generating plant—that I have been analyzing for their resistance to being broken during an earthquake or other such potentially catastrophic incident.

The design on this engineering computation paper is very clever. The obverse is unruled except for a few thin gray lines that delineate left and right margins and define several spaces across the top of the sheet for identification of the work, the date, the engineer, etc. While the exact arrangement of spaces may vary from pad manufacturer to pad manufacturer, the clear intention of the design is to block out of an 8½-by-11-inch piece of paper a 6½-by-10½-inch rectangle in which the engineer can attack the problem at hand.

The reverse of the page is covered with a light blue grid of five squares to the inch, and these show through the paper to provide an ever so faint quadrille of guide lines for sketches, equations, calculations, tables, notes, and graphs. When the page is filled and torn from the computation pad, the rectilinear grid is barely visible behind the neatly aligned rows of equations, columns of figures, and sketches of mathematical models and engineering designs. The warp and woof of the underlying analysis vanish behind the pattern the engineer has synthesized. The solution to the problem is rolled out like a fine carpet over a floor tessellated with common tile. Like tight petit point, the canvas matrix does not show through.

Engineer's computation paper is symbolic of the constant dilemma that technologists must face: the need for original thinking and innovative problem solving within the bounds of accepted practice and professional conformity—not to mention the rigors of natural law. The engineer's dilemma is not unlike that of the artist, who must constantly wrestle with the troublesome horns of substance and form. But engineers and scientists are handicapped by the additional constraint of the engineering and scientific method, which requires what amounts to universal consent among the technical community that one's design or analysis is objectively sound. No matter how irrationally conceived in a flash of genius, the work must be able to withstand the closest scrutiny of one's peers and superiors, and when an engineer knows his work will be scrutinized he more often than not sets down its rationalization on engineering paper of this or of a similar design.

Since my first days in engineering school I have been accustomed to reading the claims of the computation pad manufacturers on each fresh pad I have unwrapped from its protective cellophane. The 80 SHEETS always seemed to include the top one, on which the product itself was advertised as being LITHOGRAPHED *For True Engineering Accuracy* with NON SMEAR INK so the lines could not be smudged by perspiration. I do recall striving for accuracy while sweating through the final examinations in Resistance of Materials and Machine Design during a heat wave one May some twenty years ago, but I cannot now recall the degree of accuracy I achieved or whether the ink smeared on my computations.

Next to the engineering student's sweaty palm, the object most likely to be in contact with a computation pad in those days was a ten-inch slide rule, if it were not hanging in its scabbard from the student's belt. We were told by our professors that a good rule would last us throughout our careers, and the choice of a professional-grade "slipstick" preoccupied many a freshman for a semester or two. But by the time we were sophomores we had chosen our instruments according to their feel and the special features they possessed. It was not likely that civil, electrical, and mechanical engineering majors could comprehend, much less use, all the esoteric scales of one another's slide rules.

A wooden model made by the Post Company was gaining some popularity and challenging the long-standing favorite brand, Keuffel & Esser, during the late 1950s. I looked long and hard at the Post slide rule because I was very happy with the Post drafting instruments I had brought with me from high school, but the rule felt light in my hand and its case seemed to me to be an unattractive dark brown and bulky. There were also metal, plastic, and even circular models made by such companies as Dietzgen and Pickett, but these were preferred generally by eccentric or nonconformist engineering students. Neither my eccentricity nor my nonconformity expressed itself in my choice of slide rule, however, and I wanted none of the warping, binding, or cracking problems rumored to be had with some of the less popular designs.

My obsession with slide rules was not atypical, as the corny, chatty prose of cartoonist Don Herold indicated in a Keuffel & Esser advertisement that appeared in 1950 attests:

> Although K&E Slide Rules are the oldest slide rules made in America, there are no whiskers on 'em—except cat's whiskers (I mean, symbolizing precision).
>
> If you are an old engineer, you probably regard your K&E Slide Rule as a priceless Stradivarius. You are probably figuring on passing it down to your grandchildren.
>
> If you are a beginner, the sooner you attach yourself to an immortal K&E rule, the better.
>
> K&E Slide Rules have become accepted symbols of the engineering profession. If a photographer, illustrator or cartoonist wants to indicate that his hero is a top-flight engineer, he puts a K&E Slide Rule in his hand or in the immediate environment.
>
> Ask anybody what he knows about Keuffel & Esser, and he first starts rhapsodizing about slide rules. Slide rules and K&E are synonymous.
>
> And they're both almost as long lasting as the pyramids!
>
> Keuffel & Esser have been around since 1867 and they completed their first batch of slide rules in 1891.
>
> It is not uncommon to hear of a K&E rule which has been in service for over 50 years.
>
> I used to think a slide rule was a slide rule—just as a yard stick is a yard stick—but there is a sensational variety of 'em in the K&E line—from the simple Mannheim to the more complicated brethren, such as the Log Log Duplex Trig and Decitrig and the Log Log Duplex Vector.®
>
> Also, there are several sizes, from the handy pocket rules to the more common 10-inch, up to the 20-inch longfellows.
>
> There's no point to hitching up for life with a "second best" slide rule when you can play a Keuffel & Esser.

The slide rule I did choose was the very popular and versatile Log Log Duplex Decitrig model, which came with a thin and attractive tan hard leather case that I did not choose to equip for belt-hanging. I was always very happy with my choice, and I always believed that the slide rule that, as a symbol of the profession, was pictured on the front of the engineer's computation pads I used was just like my K&E.

It was that silent computational partner, my constant companion throughout college, graduate school, and my early engineering career, that passed through my mind when I took this engineer's pad from the laboratory supply cabinet today. For in place of the familiar slide rule, forever cocked to multiply 1.285 by something in the middle of the scale, is a stylized pocket electronic calculator displaying the number 42583.21, which is way too many digits for any ten-inch slide rule. Furthermore, the now-ubiquitous silicon chip machine calls to mind the checkbook balancer and the supermarket shopper equally with the engineer, and it is in no way a symbol of engineering the way the slide rule was for so long.

Even at the height of its popularity, the slide rule was a scientific instrument that was but a curiosity in the hands of the uninitiated. To use a slide rule properly one had to understand the principles of logarithms, those powers of ten and the interminable natural constant $e = 2.71828\ldots$ that scientists and engineers learn to manipulate as easily as pi.

Logarithms were developed in the early seventeenth century by John Napier, a Scot, as a means of reducing the occurrence of errors in the tedious calculations of products, quotients, powers, and roots that were necessary for constructing trigonometric and astronomical tables to many decimal places. With logarithms, multiplication and division were reduced to addition and subtraction, and therein lies the essential principle of the slide rule.

The component parts of a slide rule are accurately marked not in equal divisions but in divisions proportional to the logarithms of the numbers inscribed on the scales. As the scales are slid against one another, the logarithms of the numbers being multiplied or divided are added or subtracted as easily as one measures a line with a ruler. There is a certain skill necessary to interpolate accurately between markings and to add the proper number of zeroes or insert the decimal point correctly in the answer, but this facility was learned early in the engineering curriculum before a six-foot working replica of a slide rule that hung over the blackboard in many an engineering classroom.

Although the whole class of neophytes would be given the identical two numbers to multiply or divide, it was understood that they should not expect to get identical answers. Such exercises in humility and in the limitations of the human brain, hand, and eye—and the instruments of human design—taught generations of engineers the folly of seeking absolute perfection and precision.

The limitations of the slide rule were also its strengths. The absence of a decimal point meant that the engineer had always to make a quick mental calculation independent of the calculating instrument to establish whether a job required 2.35, 23.5, or 235 yards of concrete. In this way engineers learned early to develop what would become an intuitive appreciation of magnitudes.

Now the decimal point floats across the display of an "electronic slide rule" and lights among extended digits that are too often copied down without thought by the young student and practitioner. Whether the batteries of the calculator are low or the voltage high is seldom considered, and one-tenth or ten times the necessary concrete could easily be ordered without anyone being the wiser until the field engineer called for more concrete or for help. Of course there are other means of checking calculations, but it is unfortunate to be apparently losing one of the most reliable—reflection.

Although many electronic slide rules come in cases equipped with belt loops, today's engineer is more likely to carry his or her calculator in an attaché case. The once-ubiquitous slide rule tie clip, the practical limit of computational miniaturization just twenty years ago, has vanished with the engineer's tie, and the engineer's computation pad is in danger of being replaced by the magnetic tape or disk in a portable computer.

Now only gray-haired engineers work with slide rules by their computation pads. Most of the younger generation have adopted the programable calculator and personal computer as eagerly as they embraced television, and our calculations are more likely than not scribbled without discipline on the back of computer paper of no uniform size or design, and certainly not packaged with pictures of slide rules. Our handwriting, perhaps because we are intimidated by the mechanical reproducibility of the computer printout, has become untrammeled. Our designs, perhaps because of the size and abundance of the accordion-folded printout paper with its supposed justifying calculations, have become less and less substantial and substantiated. Our answers to problems, perhaps because of the idiot-savant computer and the unused slide rule in the bottom drawer, have become swollen with seemingly endless numbers with seemingly endless digits and with seemingly endless possibilities.

And when these numbers and this paper are spewed out of a giant computer by a laser printer at the rate of 1200 lines per minute, they cannot always easily be checked with a slide rule on an engineer's pad, or even with a pocket calculator. But to do so is one of the greatest challenges today's engineers face. Let us hope it is attacked with slide rules—whether wooden or electronic —a line at a time on soft-green engineer's pads before a picture of a robot operating a computer appears on the packaging.

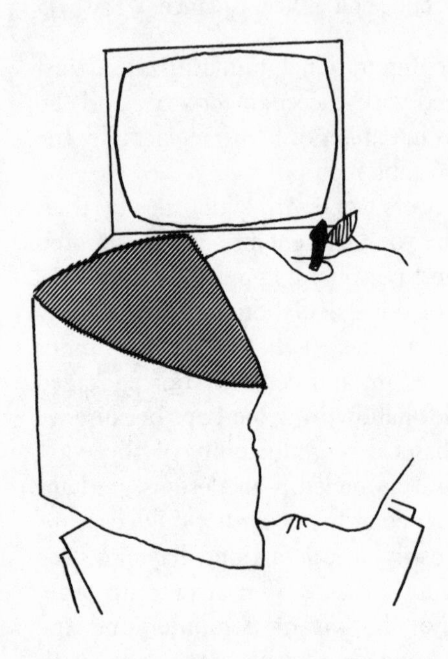

7·Logon Proceeding

My office mate has gone to lunch at the laboratory's cafeteria, but I have remained here with the cheese and fruit I have brought from home. Lunchtime is a daily respite during which I enjoy the silence of the corridor and the joys of texts without equations.

```
LOGON PROCEEDING
```

But this noon I am distracted by the low whir and the high whine of the unattended data terminal at my neighbor's desk. The drone of the cooling fan motor is punctuated at precise thirty-second intervals by the printhead skimming across the paper, leaving a trail that reads:

LOGON PROCEEDING

This portable terminal, a thirteen-pound wonder of technology, is connected to the large IBM computer two buildings away by the same telephone line over which my colleague calls Washington and his wife. The data transmission signal is whining now ever so high.

LOGON PROCEEDING

The distracting device looks like a stylish typewriter and is commonly called a TSO terminal hereabouts. The letters stand for "Time Sharing Option," a computer software system that enables several score of scientists and engineers to use remotely and almost simultaneously the same main computer.

LOGON PROCEEDING

Once an individual is "logged on," he shares the use of the main system with others who are also logged on. But although the laboratory's computer facilities operate twenty-four hours a day, seven days a week, they are inadequate. Hence one cannot log on at will.

LOGON PROCEEDING

There are many students and visiting researchers at the lab this summer, and they tax further the capabilities of the computer facilities. So many want to share time on the system that one has to queue up electronically and wait for an open region in the computer's memory bank.

LOGON PROCEEDING

To log on, one must dial one of several telephone extensions that, when they are not busy, are answered by the computer-linking, high-pitched whine. The telephone is then inserted into the "acoustic coupler muffs" at the rear of the terminal so that it and the computer can converse.

LOGON PROCEEDING

When this acoustic coupling has been made, a little green light signals the user to type in LOGON (pronounced "log on") and an identifying number and password. If the computer recognizes the identification as valid, and if the user has not overdrawn his account, the terminal session proceeds.

LOGON PROCEEDING

Depending on the number of TSO users, one may get an immediate response typed back at the rate of thirty characters per second, or one may have to wait. In any case, the data terminal will type out the user's account status and indicate that the logon is in progress.

LOGON PROCEEDING

If there are any important changes in the computer system, news of these will follow in a series of broadcast messages. The computer will also deliver any mail that has collected in its memory since this user last logged on. Mail may consist of notifications that the computer has completed previously assigned jobs for this user or may contain messages sent by other users of the system.

LOGON PROCEEDING

The "typing" done by my neighbor's terminal is performed by a rapidly moving printhead that is embedded with a matrix of tiny heating elements. The elements can be turned on and off so quickly as to change from character pattern to character pattern while the head skims across the paper, which in turn peels off a roll, flies over a "dancing roller," and lies on a platen under the printhead.

LOGON PROCEEDING

The paper is temperature sensitive, and the characters are formed by the printhead heating an array of spots in the form of a letter, numeral, punctuation mark, or mathematical symbol. If one looks closely at the output of such a terminal, one sees a text made up of little dots, much like the impression left by a cotton typewriter ribbon.

LOGON PROCEEDING

Because of the frequent long waits to get on the computer during working hours, many people carry the portable terminals home in the evening. Some have had an extra phone line installed so as not to tie up the family phone during the long terminal sessions, which take place throughout the night, when the computer's response is immediate.

LOGON PROCEEDING

My colleague has left his terminal whirring away because he hopes to use TSO as soon as he returns from lunch. Just before he left, he got a telephone line through to the computer and initiated the logon process. To avoid a dessert of LOGON PROCEEDING messages, many of those who are logged on before lunch do not log off when they leave for the cafeteria.

LOGON PROCEEDING

As the terminal executes these messages, it sounds like a very energetic woodpecker: it pecks out thirty characters in a second and returns to the margin as the log of paper unrolls another line. The thud at each return of the printhead sounds like a stricken bird falling to the ground.

LOGON PROCEEDING

When a user has gained access to a TSO region on the computer, he and the machine will alternate typing. A staccato dialogue in the laboratory's own Speakeasy or some other appropriate language will take place, with the printhead returning to the margin within two hundred milliseconds after each command or response.

LOGON PROCEEDING

A user may want to gain access to TSO to inquire after the status of a big job that had been submitted to the computer earlier in the day, or even the day before. It is not first come, first served on the computer, and the exact order in which jobs are executed follows a complex and seemingly whimsical formula.

LOGON PROCEEDING

Generally, small jobs that require only a few seconds of computer processor time are executed before larger jobs that may require minutes, even though those larger jobs may have been waiting in the queue for hours. One way of getting a job to run sooner is to raise its priority, thus multiplying its cost by a factor of two or three. Priorities may be raised at a remote terminal.

LOGON PROCEEDING

When one is dealing with complex physical problems formulated in complex computer programs known as codes, it is often necessary to get as rapid as possible a turnaround on the computer. The next step in the analysis often depends greatly on the results of the last. Waiting for the computer to execute a big job can be a frustrating experience.

LOGON PROCEEDING

The computer user waiting for output is like the writer waiting for the mail, the photographer waiting for his film to develop, the jockey waiting for the official photo, the crap shooter waiting for the point, the farmer for the crops to come in. There is an anxiety about failure and a hope for success.

LOGON PROCEEDING

My neighbor's terminal now has about eight inches of reassurances above its keyboard. He should be returning from lunch shortly, but the computer can't know that. Only when there is an available space for another user will a characteristically short tattoo spell out the word my colleague hopes to come back to—

READY

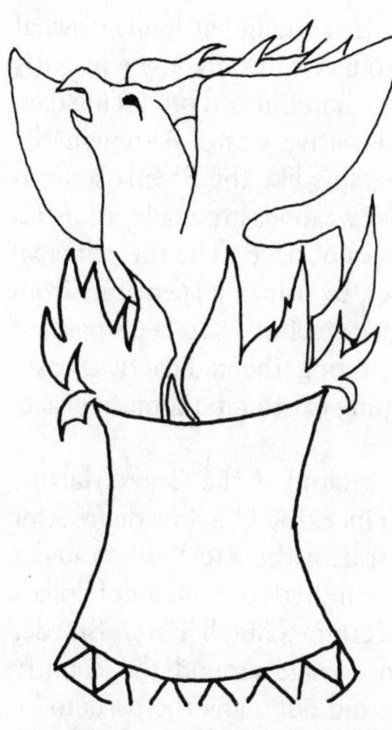

8·How Poetry Breeds Reactors

The technological superiority of the United States—the country that put men on the moon and brought them back—has been threatened by more than foreign steel and automobiles. France's nuclear breeder reactor program is the envy of nuclear engineers and physicists throughout the world, though their own national pride allows only an infrequent acknowledgement of the fact.

Phénix, a liquid-metal fast breeder reactor, has been generating electricity so reliably since it went critical in 1974 that France, in cooperation with Italy, West Germany, Belgium, and the Netherlands, is well on its way to completing a larger version of the same concept. SuperPhénix, which in 1985 was in the final stages of testing before reaching criticality, will have the capability to provide 1240 megawatts of electricity, about five times as much power as Phénix puts out, while converting low-grade uranium into plutonium to fuel future reactors in Western Europe.

The power level of SuperPhénix is at the upper limits of what conventional U.S. reactors now produce not only less efficiently but also while consuming high-grade uranium and producing dangerous and presently unusable radioactive waste. Although the plutonium produced by breeder reactors like the Phénix series is the very toxic stuff of which nuclear weapons are made, France's program demonstrates that this does not have to be the principal or final use of the substance. In fact, by utilizing spent fuel from conventional nuclear plants, breeder reactors have been pointed to as a means not only of ameliorating the radioactive waste storage problem but also of consuming existing plutonium peacefully.

As SuperPhénix rose out of the ground at the Creys-Malville construction site in southeastern France, the U.S. breeder reactor program was, as it has been since at least the late 1970s, a major technological embarrassment, with hundreds of millions of dollars worth of components for the uncertain Clinch River Breeder Reactor Project fabricated but in storage around the country because policy and licensing delays did not allow the parts to be assembled or even collected at the Tennessee Valley site.

Progress reports on the Clinch River project were conspicuously absent from the *Transactions* of a 1980 international conference on world nuclear energy, sponsored jointly by the American and European Nuclear Societies, while the French experience provided a leitmotif throughout the meeting. Successful Phénix was an especially conspicuous specter at the closing plenary session in which the week's accomplishments and their implications for the future of world nuclear energy were summarized not by a representative of the U.S. Department of Energy, headquartered right there in Washington, D.C., where the meeting was held, but by a director of Électricité de France, that country's state-run power-generating authority.

There are many and complex technical, political, and cultural reasons for the superiority of the French in the relatively narrow area of expertise comprising breeder reactor technology, but their marching forward, and to a different drummer, while others mark time in step, if they are not indeed retreating, can be more easily grasped in terms of the stuff of poetry—language and words and the naming of things—than in terms of the equations and data of science and engineering.

The goals of the French and the floundering American breeder reactor programs were fundamentally the same. Each country wanted to develop and demonstrate the commercial and technological reliability of large breeder reactors in order to use nuclear fuel more efficiently, thus avoiding the use of foreign oil and guaranteeing energy independence for themselves. However, the success of a breeder reactor demonstration project, like that of the Manhattan Project and Project Apollo, which were of comparable ambition and innovation, depends greatly on *esprit de corps,* that nontechnical intangible that unites myriad individuals with a common purpose and enables them collectively to accomplish goals beyond the simple sum of their capabilities. Yet this necessary ingredient, which is universally uttered in the French idiom, does not appear to translate well in word or in deed from the French breeder program.

Their language itself seems to have set the French technologists apart from the rest of the world and given them both the *esprit* and *individualité* to build a breeder program second to none. Although the official languages of many international technical conferences are English, French, German, and sometimes the language of the host country, by and large it is English (more precisely American technical English) that dominates the proceedings. Only the French, who always seem to speak eloquently the language of their poets, can be consistently expected to deliver their papers in a language other than English, which everyone else, including Japanese, Germans, Italians, and even Americans, often butchers into a nuclear lingua franca.

The tenacity of the French for their language pervades the

international nuclear reactor research-and-development community. When a seminar on safety aspects of liquid-metal fast breeder reactors was being organized in the late 1970s in connection with one of the biennial international conferences on structural mechanics in reactor technology, the American coordinator invited, by writing letters in English, a score of panelists from a half dozen countries to participate. Not only did the French turn down the invitation, presumably because the operators of Phénix could learn little from such a seminar, but also they declined in colloquial French. Representatives of other countries accepted gladly or declined apologetically in their best poor English.

The programs of such meetings have always been printed in English, even though West Berlin has been the generous host of three of the eight conferences held to date. The announcement for the sixth conference, which was held in Paris, was printed in French and English, however, with the host country's language primary. This bilingual precursor was in sharp contrast to the final program of the fifth meeting, for example, in which even the welcoming address of the mayor of Berlin was printed not in German but in English.

This determination of the French to resist international conformity in language extends into the hard technological arena of breeder reactor design—and here even in English a curiosity of words makes the differences accessible to the layman. Basically there is a choice between two fundamental arrangements of the components of a liquid-metal fast breeder reactor plant: the essential transfer of heat from the plutonium core to the steam generator can take place either inside or outside a reactor vessel that is filled with the liquid sodium metal. This vessel provides the primary containment for the reactor's energy and protection against undesirable radiation to the exterior.

If the pumps and heat exchangers that carry thermal energy away from the fissioning nuclear core are immersed in the sodium within the primary vessel, the reactor is known as a *pool* design; if they are located outside the vessel and connected through a problematic and vulnerable network of pipes, the reactor is a *loop* design. The fact that the words *pool* and *loop* are not only anagrams but also reversals of each other underscores the fact that the two reactor designs, while having the same major components in different arrangements, are as different in their technological connotations as are two words that share letters of the alphabet.

The French reactor Phénix has demonstrated the soundness of the pool design, and it is being employed in the SuperPhénix also. The Americans, on the other hand, having opted well over a decade ago to develop the loop design, may never complete the demonstration project at Clinch River, which became mired in political controversy and was questioned even on purely technical grounds as an idea whose time had passed.

The pool design is not alien to the American breeder reactor program, and there was in fact for many years a very determined experimental breeder reactor of the pool type operating at the Idaho site of Argonne National Laboratory. Although recently classified as an experimental rather than a demonstration reactor, the original intention of this successor to the first experimental breeder reactor was to demonstrate the technical, if not the economical, viability of the breeder concept. Experimental Breeder Reactor II, which went critical in 1965, provided indispensable data on the behavior of fuel and engineering materials in a deleterious fast-neutron environment while generating enough power to supply the electrical needs of the reactor operations building itself.

This admirable record had to come to an end, however, since it was in the American breeder reactor program plan to convert the 30-megawatt pool system to a safety test reactor when the Fast Flux Test Facility, a loop design known as FFTF, finally went critical. The fact that the criticality of FFTF, which generates no power, was well over five years behind schedule is one of the many embarrassments American nuclear scientists and engineers have had to suffer, usually in silence.

During the 1976 presidential campaign, Jimmy Carter, himself an erstwhile member of this country's (and Admiral Hyman Rickover's) crack nuclear navy, acknowledged in an interview that other countries' breeder reactor programs were further along technologically than ours and that, if elected, he would push for international cooperation for the mutual benefit of all. After the election, President Carter made clear his administration's low opinion of the country's demonstration breeder reactor project at Clinch River, and the future of the project was a political football for years.

During that time opportunistic, frustrated, insecure, impatient, and simply bored nuclear engineers and physicists quit the American breeder program by the scores, leaving behind a subcritical reactor corps devoid of *esprit* but haunted by the spirits of its youth. As SuperPhénix is a scaled-up version of Phénix, so the Clinch River reactor is of FFTF, and therein lay the fear for some. As much as the French expected a dependable second-generation demonstration reactor plant to go critical in a few years, so the Americans, wish as they might that it be otherwise, had reason to be anxious that Clinch River, only a first-generation power-producing plant comparable to Phénix and not SuperPhénix, would also be plagued by labor, political, and technological delays well beyond the 1980 start-up date projected a decade earlier.

Shakespeare asked, "What's in a name?" and answered, but not in so few words, "Not much." Poets, like scientists, think and argue not by name but by analogy and metaphor, and Shakespeare, after handing his readers a rose to sniff, gets them to think by example that what someone or something is called is of no essential consequence. Yet a rose is a given of nature, while men and breeder reactors are shaped from their conception by their forebears and designers—and by their given and inherited names.

These names influence an individual's image of himself or the collective image of the *raison d'être* of a research-and-development team. And certainly names influence our perception of the shape of things to come, especially when they are nuclear breeder reactors with gestation periods decades long.

Nuclear power plants in the United States are generally named, if not after a local hero or civic booster, after their geographic location, and a roster of existing plants reads like a gazetteer of battles from an American history book: Indian Point, Browns Ferry, Turkey Point, Millstone, Diablo Canyon, Three Mile Island. And the pejorative connotations of many of the names do not suggest victory in the battle or in the war.

One exception to the naming of U.S. nuclear power plants was the Enrico Fermi sodium-cooled fast reactor, which operated just outside Detroit in the mid-1960s, until its core suffered extensive fuel damage and melting when a foreign object obstructed the coolant flow, a harbinger of the setbacks the U.S. breeder program would suffer. And when a book was written about the incident, it asked not what went wrong with an American reactor named after the Italian discoverer of the new world of nuclear power, but, instead, how we almost lost a French-named city on the St. Lawrence Seaway.

Nuclear power plants in France are also identified by their geographical location, but French breeder reactors proper have designations that have all the poetry that the names of U.S. reactors lack: the liquid-metal fast breeder reactor concept was demonstrated in France to be viable by a fast reactor named Rapsodie. Rapsodie began operating only in 1967, the twenty-fifth anniversary of the achievement of the first sustained nuclear chain reaction by Fermi and his team at the University of Chicago.

A name like "Rapsodie" softens the impact of hard technology and humanizes it in ways that Experimental Breeder Reactor II, let alone its nickname, EBR-II, cannot. "Rapsodie" evokes images of a symphony of atoms splitting in concert to a score as grand as the music of the spheres. To the erudite who know that a rhapsody in Greek antiquity was a portion of an epic poem adapted for recitation, a breeder reactor program that uses Rapsodie to exhibit the feasibility of an enormous long-range research-and-development program to make France energy-independent has the ring of classical scholarship and not of cold technology.

After Rapsodie accomplished its objective it was converted to an experimental reactor, but its name remained unchanged. The successor to Rapsodie in the French breeder program is Phénix, and its name could not be more apt. The phoenix in Egyptian religion is the avian embodiment of the sun god that consumes itself in fire and rises rejuvenated out of its own ashes. There would seem to be no better metaphor for the breeder reactor, whose fission process was once thought to be like the sun's itself and whose burning core transmutes spent uranium ashes into plutonium capable of fueling the next cycle.

Unlike the French breeder or even America's own manned space-flight program, which drew some of its strength and intelligence from classical allusion, the U.S. liquid-metal fast breeder reactor effort has confounded tongue and ear almost from the start with its own unpronounceable abbreviation, LMFBR, and the uninspired names of its milestones. Except for the whimsically named early fast reactor Clementine, the U.S. program has christened its lumbering past and future generations: Experimental Breeder Reactor, Fast Flux Test Facility, Clinch River Breeder Reactor, Prototype Large Breeder Reactor, and, finally, Commercial Breeder Reactor. No poetry is lost, or gained, when these are condensed to EBR, FFTF, CRBR, PLBR, and CBR, unpronounceable code names that are not even true or clever acronyms. Even Canada, in its independent pursuit of the deuterium, or heavy-water, moderated reactor, spurred on its own research and development team and instilled confidence in prospective customers by contracting its Canadian-deuterium concept not into CDC or some other cold string of letters but into the positive and self-confident nickname, CANDU.

Why a technological community that concealed the early stages of atomic energy under the code-name Manhattan Project, dubbed the first atomic bombs the innocuous-sounding Little Boy, Thin Man, and Fat Man, christened a nuclear navy fleet Trident, gave us the quark with its properties of charm and color, and wanted to build a cyclotron named Isabelle, could not come up with better names for its LMFBR program and its progeny may forever remain one of the greatest mysteries in the history of technology.

The somber and uninspired behavior of the U.S. nuclear technologists and managers in naming the products of decades of effort and billions of dollars of funds cannot have helped the uncertain future of the LMFBR program. It is a long way from the playful attitude on the squash court at the University of Chicago where the first nuclear pile was constructed in 1942. There, where forty-three scientists and engineers celebrated over a bottle of Chianti when a control rod was pulled out to throttle the world into the nuclear age, the main control rod was not named MCR but Zip. It is the lack of "zip"—an American ingredient as positive as *esprit de corps*—and the general absence of poetry in the U.S. breeder program that have no doubt contributed to its past and present lethargy, if not its downright demise.

At the University

9 · You Can't Tell an Engineer by His Slide Rule

Engineering students were once portrayed on a "Saturday Night Live" television show as gauche white males whose sport shirt pockets were bulging with pencils, pens, and other implements of their profession. In the skit, which took the form of a fashion show, one model displayed a new plastic pocket that adhered to his bare chest so that he might carry his paraphernalia to the beach, while the other model posed in a flannel shirt with a slide rule dangling from his belt and an electronic calculator in his outstretched hand.

Except for the now-ubiquitous calculator, which is as symbolic of modern Everyman as of the technologist, the skit more successfully dated its writers than it satirized today's engineer. Not only is the redundant slide rule obsolete, but so is the stereotyped engineering student, and one must look beyond outmoded badges and clichés like pocket protectors and flannel shirts to identify who is and who is not studying engineering at our universities these days.

There are many new characteristics to isolate in caricature, and they can be found in the controlled-circulation magazines that portray the engineer as he or she appears to the advertiser. Advertising is the most faithful and biting medium of satire, for it is always looking ahead to a market and anticipating, if not forming, the future foibles of its audience. The measure of success is not the transient needle of a laugh meter but the bottom line in an annual report.

The companies that advertise themselves in magazines like *Graduating Engineer* are trying to sell the company to the potential employee. Thus they portray the engineer as they think the young engineer wants to see himself or herself in the context of the corporate structure. A common tableau in these advertisements is a group of three engineers studying a blueprint or looking at a computer terminal screen. In an uncanny number of instances, only one of the three engineers is a white male. The others are usually a black male and a white female, but racial and ethnic types are often so blurred that the models might also be of Hispanic or Asian descent. What is incontrovertible is that the engineers do not have pencil cases stuffed in their shirts, and they are dressed in a spectrum of colors and styles ranging, at least in the mid-to-late 1970s, from tacky blue leisure suits to preppy pink Fair Isle sweaters.

Although the tableau tells a white lie by underrepresenting the white male, who is certainly still in the majority among engineering students, the idea conveyed by the heterogeneous group is a valid and fair one. Engineering in this country is an egalitarian discipline, and its students come from the immigrant-hopefuls and first-generation American families as well as from families whose pedigrees reach back to the Revolutionary War and qualify them for membership in the exclusive Society of the Cincinnati. Thus while some engineering students see the profession as an opportunity to get a leg up on the social ladder, others view an engineering education as the firm footing upon which they will set the corporate ladder in preparation for the climb to the high-level positions their fathers and grandfathers, engineers before them, already hold in major corporations.

Since the population of current engineering students is nowhere near so homogeneous or recognizable as "Saturday Night Live" suggested, engineering is not the object of campus jest or protest that it was when the show's writers were probably in school. For example, at Duke, which had long been dominated by the liberal arts and sciences of its Trinity College, unexpectedly large numbers of students transferring from liberal arts into engineering swelled class sizes and taxed facilities in the early 1980s to the point where such transfers became very difficult for the engineering school to accommodate and were, therefore, not encouraged. (Because it is more difficult to gain direct admission to the School of Engineering, it had become a common strategy to seek admission to Trinity while premeditating a transfer into Engineering.) Engineering is no longer considered the major of oddballs and warmongers but is seen as a source of solutions to technological, environmental, and social ills—as well as a virtual guarantee of a job upon graduation.

The situation is similar throughout the country. Undergraduate engineering enrollments have returned to and exceeded their peaks of the late 1960s, from which they had plummeted. The lean years of the 1970s have left their mark on engineering education, however, and engineering laboratories and faculties are in need of modernization. Much equipment is as obsolete as the slide rule, and faculties are understaffed. And ironically, now that America's engineering schools are overcrowded, understaffed, and poorly equipped, there seems to be an insatiable demand for their graduates. A bachelor's degree in engineering can command a starting salary in the high twenties, well beyond the salaries of young assistant professors who spent four, five, or six years beyond their own bachelor's degrees to earn engineering doctorates. This kind of inequity has precipitated a series of problems that have impeded engineering education at a time when politicians and educators, corporations and foundations, are expressing concern over America's ability to remain technologically fit in an increasingly competitive world.

The outlook had become so bleak that in February 1980 President Carter posed formal questions on the condition of science and engineering education in this country to both the director of the National Science Foundation and the secretary of education. They responded with a co-authored government report, *Science and Engineering Education for the 1980s and Beyond*, issued jointly by the two agencies in October 1980. Released at the height of the presidential campaign, the report received little notice then and it has since remained a curiously unreferenced and generally unacknowledged document, even though the conditions it describes still by and large prevail and its message is still current at mid-decade and is reiterated in the 1985 report of the National Research Council, *Engineering Education and Practice in the United States: Foundations of Our Techno-Economic Future.*

The report to President Carter unequivocally documents and warns about the worsening U.S. technological prognosis. Among the principal findings of the appraisal of the adequacy of science and engineering education for the country's long-term needs are: (1) that there are shortages of most types of engineers at all degree levels; (2) that the adequacy of the numbers of Ph.D. engineers in 1990 is "problematic"; (3) that there is a shortage of high-quality engineering faculty; (4) that minorities are underrepresented on engineering faculties; (5) that the cost of maintaining existing laboratory equipment is adversely affecting the education of advanced graduate students; and (6) that many newly graduated engineers are not adequately trained in state-of-the-art techniques because engineering schools lack sufficient resources to modernize teaching facilities and equipment. (Additional findings of the study pertain to the closely related problem of secondary school science and mathematics education and its relation to the university problems and to the general problem of what has come to be known as "technological literacy.")

Not only did the first Reagan administration disregard the alarming report to its predecessor, it has made budgetary decisions that would appear to aggravate conditions. Science and engineering education programs were among the most drastically cut segments of the National Science Foundation's budget in the early 1980s. And revisions of research budgets left many engineering faculty and their graduate students, who had laid out elaborate research programs already approved by the Department of Energy and other agencies, in limbo or worse, waiting for checks from Washington that never came.

When young engineers with hot bachelor's degrees are being wooed by affluent industries, it is not surprising that they are not attracted to graduate schools with dilapidated labs and iffy grants. And a graduate faculty with fewer graduate students works not less but more, attempts to recruit more students from fewer applicants, writes more unfunded proposals for fewer funds, does more research with fewer assistants, teaches more classes with fewer colleagues.

The diminishing pool of good American applicants to graduate study in engineering has led to a steadily increasing number of engineering graduate students with bachelor's degrees from foreign institutions. Recent figures indicate that about half of the total number of engineering doctorates awarded in this country now go to foreign students. This has created problems on two fronts. Not only is there a question of unintentionally exporting important advanced technological experience to competing countries, but there is also a problem of replenishing the maturing engineering faculties of our own country with young assistant professors with a command of English as well as a command of technology.

Some observers feel that the reluctance of young American engineers to earn doctorates and enter the teaching profession is only indirectly a matter of money. They believe the problem rests in the lack of role models among engineering faculty, and this view is supported by the facts. Certainly there is a paucity of members of racial minority groups and women teaching engineering classes comprising increasing numbers of students who are not white males. But, perhaps more importantly, the lot of today's (albeit white male) engineering professor—teaching enlarged classes with poor equipment in crowded classrooms, serving on search committees for potential faculty who do not seem to exist, and looking for a summer's oasis of research support that appears to be as palpable as a Cheshire cat—often appears to students to be as harried or more so than the lives of their businessman fathers.

Undergraduates see that an element of joy is absent from teaching for professors who must rush from class to committee meeting to oral exam to laboratory to class. What could recommend this life to young men and women who will be asked to spend several years on peanuts and loans—in order to earn a doctorate in order to qualify for an underpaid position in order to solicit support from undergraduate classmates who have climbed the corporate ladder in private industry? Although there may be a financial impediment to attracting talented people to engineering teaching, there is also this something more intangible.

Some of these problems are not unique to engineering, of course. I see and hear of similar situations among my colleagues in the physical and social sciences, the arts and the humanities. We sit together on committees for the Program in Science, Technology, and Human Values, and we discuss the problem of disciplinary progress over interdisciplinary lunches. The ebb and flow of attendance at these sessions attests to the revolving obligations of us all, even as we count on these chance encounters to exchange views before the next committee meeting on curriculum reform or faculty interaction.

Nevertheless, these interactions help allay the paranoia that we are each of us members of a most singular and beleaguered group. It seems that a fresh supply of humanities faculty in the 1990s is also problematic, but not because of a surfeit of openings now, as in engineering, but because of the past two decades' dearth, which has made students find, for opposite reasons, a humanities Ph.D. a risky investment of time, energy, and hope. As positions on humanities faculties have become more scarce, the ranks of the independent scholar who continues to pursue research even without an academic base have multiplied. On engineering faculties, the adjunct professor, who continues to be a worldly engineer, has of necessity also become an increasingly common species. But while adjunct professors and independent scholars contribute in unique ways to their respective professions, they cannot be expected to ensure the continuity upon which academic programs and universities themselves depend. That requires a constant supply of and demand for doctoral candidates of all colors and both sexes in all fields. There is a heterogeneous mix of potential faculty among each year's graduating class of young engineers, but until economic conditions or the lifestyle of present faculty changes significantly, the racial and sexual composition of engineering faculty, if not its trappings, may be accurately portrayed by satirists who came of age in the 1960s.

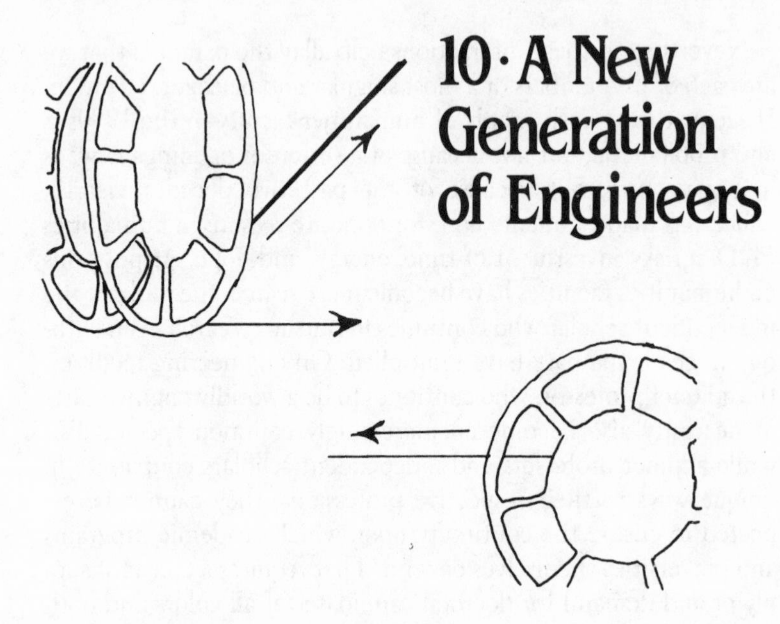

10 · A New Generation of Engineers

The semiannual task of looking over class rosters and assigning final grades has never been an easy one for me. The alphabetized lists of names still evoke memories of my own student days, and I remember all too well the spectrum of feelings—from elation through perplexity to despair—that the final reckoning was capable of producing. Some things never change; other things do.

Twenty years ago I was an engineering student; today I am a teacher of another generation of engineers. And the latest class roster reminds me of changes the two decades have brought. My own classmates all had names like Brian, Joe, and Mike. Today my grade sheet lists their namesakes, but it also includes the likes of Karen, Belinda, Audrey, and other unmistakably feminine names. In fact, there is a one-to-three female-to-male ratio among engineering students now at Duke, and nationwide more than fifteen percent of engineering students are female.

As recently as ten years ago, a woman engineering student was still a rarity, although women were then being actively recruited to fill the empty classroom seats left in the wake of a drop in (male) enrollment in engineering. Now engineering enrollment is taxing the facilities and faculties of universities, and young women and men compete on the basis of merit and aptitude for seats in crowded classrooms.

It should really come as no surprise that young women and young men can have an equal aptitude for engineering. After basic science and mathematics, the first courses in an engineering curriculum are the so-called engineering sciences, really extensions of physics, chemistry, and calculus to the fabricated universe of engineering rather than the given universe of what used to be called natural philosophy. The engineering sciences typically do not deal with electrons, neutrons, and protons in the forms of abstract volumes, energies, and quanta, or in the form of submicroscopically distant atoms and molecules or astronomically distant planets and solar systems. Rather, the sophomore and junior years of engineering are filled with courses that deal with the palpable objects of experience: balls and wheels, bricks and sticks, strings and things so common and familiar to everyone as to become as fascinating and full of unsuspected mystery under an engineer's scrutiny as the flora and fauna of his own backyard can be to a naturalist. And the engineering sciences of fluid mechanics and thermodynamics treat the four common elements of air, fire, water, and earth as sources and sinks of energy, as aids and impediments to motion, and as the stuff of heating and air conditioning in ways unthinkable to the Greek philosophers.

Neither boys nor girls have an exclusive claim to the objects of engineering science, and all children employ them in play. A ball is as essential to a game of jacks as to a game of catch; wheels behave no differently on a bicycle with or without a crossbar; sticks and stones break everyone's bones; and no young boy's game uses a rope in as complicated a pattern of motion and equilibrium as does jump rope in all its variations. While it is commonly held that if not the objects then the activities of boys—building and taking apart, making and breaking things—are training ground for engineering, the activities of girls are no less so.

The archetypal girl who bakes cookies and sews dresses is in fact doing quintessential engineering. She is synthesizing. Out of a cupboard of ingredients she learns to produce by mechanical mixing, controlled heating, and human judgment a cookie that with practice is as good as Mom's. Out of limp fabric and weak thread she assembles a durable structure of form and function that is the essence of civilization. And in doing these things every young girl is designing processes and objects more successfully than most boys ever will—even after they become engineers. While Ruth Schwartz Cowan argues convincingly in her book, *More Work for Mother,* that technological advances have freed men but not women from the drudgery of running the household they once toiled in together, her book also makes incontrovertibly clear that the running of a household is in itself a great technological process. From this point of view it might even be argued that girls who emulate their domestically creative mothers should make even *better* engineers than the boys who follow in the footsteps of their fathers as mere managers of processes, structures, and implements created by others.

Whatever the long-term implications of the changing roles of men and women in the home, their coequal roles as engineering students are already evident. Not only have women become integrated into engineering classes, but also they have done so with statistical anonymity. In virtually every course I have taught in the past few years, the grades of the men and women together have been normally distributed with no apparent sex skew. The top student's name has been Sherri as often as it has been Tom, Dick, or Harry. Indeed, like top students of all generations, she has often been in a class by herself when it comes to resourcefulness, and she has lost none of her femininity in her quest for excellence and in her display of ingenuity.

Once a student in my course in dynamics produced in class a bobbin from her sewing machine to demonstrate the solution to a homework problem that had had the rest of the class, male and female, flip-flopping over the answer for several days. The problem is a classic one in rigid-body kinematics: Will a spool or reel of cable sitting on the ground roll toward or away from you if you pull on the end of the cable coming off the bottom of the spool? Some textbooks make the problem concrete by describing the cable as telephone wire and the spool as the big wooden kind that students are accustomed to turn on its side and use as a coffee table. Almost all textbooks show by means of a diagram the spool ready to roll on its circular edges as the cable is stretched out parallel to the street on which the telephone linemen might be working. More recent texts might even describe the problem in the context of laying fiber-optic cable, but it is seldom explained why the spool is on the ground rather than being in its customary position on an axle mounted on a trailer. (The solution to the problem should itself raise that question.)

Such problems, artificial as they may be, are ubiquitous in the textbooks used in sophomore and junior engineering courses. These are the courses where analysis as opposed to design is taught, and the situations to be analyzed are often necessarily artificial because real ones tend to be too complex to illustrate the single principles under discussion. Thus spools no lineman ever pulls, idealized balls colliding in the absence of gravity, and spinning records traversed by drunken insects are juxtaposed on examinations in bold disregard of reality but in anticipation of the varied problems an engineer will face during his or her career. Yet it is the principles of rolling, impulse, momentum, and kinematics abstracted in such problems that students are expected to synthesize in senior design courses, and it is this aspect of an engineering curriculum that makes it different from that of any of the pure sciences. It is exactly such unreal situations that become real in the open cargo bay of a space shuttle when a new satellite is to be launched or an old one captured and repaired. And if males have more experience in that environment than females it is for the same reasons that males dominated engineering for so long —they were simply there in greater numbers.

We have all, boys and girls alike, watched our mothers use their sewing machines, but none of the hundreds of young men or women in the scores of dynamics classes I have taught equated playing with mother's notions with pulling the end of a cable from a spool—until Sherri did. When I asked the class which of them thought the spool would roll away from and which toward the hand pulling the cable, I got the response that I had learned to expect—the class was about equally sure and equally unsure that the spool would roll to the right and to the left. When I asked students to support their contentions, I also got the expected responses. Those who had the right answer had all the wrong reasons, and those who had the wrong answer had some of the right reasons. Many students are quite sure of their answer, but it is the rare student indeed who has the right answer and the right reasoning and the right words to express it or apparatus to demonstrate it. Thus when Sherri produced from her bag the bobbin that could provide the incontrovertible evidence of a real experiment, the whole class was as stunned at the brilliance of the simple action as they were with the magic of the motion. When Sherri placed her bobbin on her desk and caused the thread to be wound up as she seemed to pull the thread off, the class was as intrigued as if they had just watched a game of three-card monte on Broadway. The bobbin was pulled this way and that way on the desk to show that it was not simply a sleight of hand trick assisted by gravity. They saw it but they couldn't believe it.

The conventional wisdom has long had it that the experience boys have with blocks and Erector Sets and, later, automobiles, gives them an advantage and an inclination toward things mechanical, if there is not already an innate preference for all things technological. Yet the traditional experience of girls with the chemistry of cooking, the dynamics of rope jumping, and the mechanics of sewing can be argued to give them at least the same advantage. And the synthesis required to cook a meal or sew a dress can arguably be said to be more akin to engineering than anything any boy does. In fact, even the boy who takes a clock or a radio or an automobile apart and puts it back together is merely reassembling what he disassembled. The girl who concocts a new dish or designs a new dress is really doing engineering. And what more ingenious mechanical device is there than the doll that closes her eyes and cries "Maaaaa" when put down in her bed? No, it has not been the toys or play of childhood but perhaps the constraints of motherhood and the prejudices of adulthood that for so long kept young women from studying to be engineers.

The combination of the women's liberation movement and low male enrollments in the 1970s has brought American engineering around to being the asexual field that it should be. Women now hold their share of elected student offices in our school of engineering, and they capture a representative proportion of honors at graduation each year. However, since the influx of women into engineering has occurred in less than a generation, there is one area in which women are still underrepresented.

The engineering faculty at Duke, like engineering faculties at other institutions, has a paucity of women among its ranks. The vast majority of young female engineers, like their male counterparts, have been attracted to industry right after receiving their bachelor's degrees, and they have not pursued the advanced academic training required for a teaching career. Hence the young female engineering student is generally without a role model on the engineering faculty.

In time this situation should change. The more academically inclined women engineers may begin to return to graduate school, as many male engineers have done. And they should be better teachers for their experience in industry, where problems are more real and complex than in textbooks or even in the home.

Thus they will bring a further practical perspective to engineering education. And as women assume positions in the professorial ranks in engineering schools, their assimilation into the profession will be complete. It may take several generations, but someday the grade sheets for engineering courses may contain a Jane for every John, and the instructor may as likely be a woman as a man. But even then those lists of students will remind the instructor of his or her own student days, and the task of assigning final grades will be just as difficult as it is now.

11·The Quiet Radicals

Engineers were able to feel somewhat vindicated in the early 1980s by the continuing return of radical celebrities of the 1960s to the mainstream of American society. First Bob Dylan built a million-dollar house in California. Then Rennie Davis became an insurance agent. Eldridge Cleaver supported Ronald Reagan for president. And Jerry Rubin became an investment banker on Wall Street. These same names—personifying alienation, anarchy, and alternative lifestyles—were frequently invoked during the campus unrest of the Vietnam era in arguments *against* business as usual.

Not a few engineering professors and students did a great deal of soul-searching over how to respond to radical demands for their participation in moratoriums, protest marches, and the like—actions that often were directed against engineering as a symbol of the technological establishment that allegedly was promoting the war in Southeast Asia. Conscientious engineers knew that as a profession theirs was no more responsible for Vietnam atrocities than poets as a class were responsible for the rhymes and jingles of mind-numbing television commercials. Yet the 1960s were intolerant times, and to be an engineer then was to be held responsible, or at least accountable, by the radical left.

Some engineering students and professors did take coldly unsympathetic and ultraconservative stands, which reinforced stereotypes, but by and large the technical community, like the general populace, could not be simplistically divided into hawks and doves. All fields attract a lot of queer birds and odd ducks who simply do not fly with the flock or take their places in the vee of migration, and engineering is certainly no exception. Although one might claim to identify a certain class of engineers by their haircuts and their ties, another by their deportment and demeanor, and still another by their taciturn and quiet manner, contemporary engineers are in fact nonconformists—as individual privately as celebrities strive to appear to be publicly. What any one engineer thought of Vietnam and related matters may not have been shared with or by a single colleague.

The tirades of the vocal, if not always articulate, radicals fell on many an undeserving ear during the late 1960s. A disrupted class in the rudiments of engineering mechanics more often than not contained students with majors as diverse as aerospace, biomedical, chemical, civil, electrical, mechanical, nuclear, and petroleum engineering, and not one of these disciplines fundamentally promotes war or genocide over the peaceful applications of engineering science.

Furthermore, it was unlikely that many students in those times could have said with certainty how exactly they would apply those rudiments a few years hence. That would depend on the job market at interview time and on a host of other external factors more under the control of masters of business administration than of bachelors of engineering. Internal factors, including altruism and conscience, both expanded and limited career horizons, and many an engineering freshman knew he would never work for the defense industry or that he would join the Peace Corps four years hence. But the 1970s brought such a decline in engineering opportunities generally that only those students with a family business to go home to, or with a singular vocational goal toward which they had long strived or schemed, could speculate about life after graduation.

Regardless of their career goals or hopes, what attracted many an engineering student to his major was not the problems with which the profession deals but the common method by which it attacks those problems, not the ends but the means. The methods of rational problem solving were hinted at in high-school mathematics and science. An affinity and aptitude for those subjects, often coupled with socioeconomic backgrounds that rendered the financial security of utilitarian college degrees highly desirable, sent vastly more young people to engineering schools in the post-Sputnik decade than did any wish to build bombs. Demands by campus radicals that engineering students rationalize or justify their curricula were thus confusing, not to mention irrational, unjustified, and ironic.

The engineering curricula, then as now, taught the engineering method as disinterestedly as the humanities curricula used to teach the methods of rhetoric. Can one imagine an English professor being accountable for students who might later write racist propaganda or genocidal tracts? Is a philosophy professor responsible for honing the mind of a prelaw student who might graduate to defend some unscrupulous power broker?

Still, some of the demands for moral accountability and relevance were justified. The technological know-how that had landed men on the moon, and brought them back, had not been as vigorously or universally applied to expanding the itineraries of physically handicapped earthlings. The global networks for energy distribution were not matched by humanitarian networks. Examples were legion, and the engineer knew them as well as anyone.

But engineering per se, if not always responsive, was not irrelevant. The fundamentals of wheelchair design rest upon the same basic principles as do those of lunar vehicles, and the methods of food-processing engineering are comparable to those of weapon, chemical-process, and power-plant design. Relevance is a question of ends, not of means, and to have lashed out against engineering education was to have put the heart before the course.

Now times have changed, and engineering is as wholesome as vegetarianism. Even Jerry Rubin invoked it to rationalize his decision to join the establishment, and he described himself as "a financial talent scout seeking out new companies, particularly in the area of energy conservation and development." Having met many enterprising people on the lecture circuit he traveled in the wake of his radical celebrity, he professed a wish to "connect them to the world of finance and positive investment." The 1980s were to be a problem-solving decade, according to Rubin.

To engineers, all decades have been, are, and will be problem-solving ones. These *quiet* radicals constantly work at getting to the root of problems, and if their efforts are sometimes misdirected, it is more often the direction rather than the method that deserves the blame. The engineering method is potential; the financial is actual. Not a single nuclear plant could have been built without the enormous and necessary capital that investors were willing to risk, probably more for financial than for societal gain. Jerry Rubin has surely sought alternatives to nuclear power, but his sources of capital no doubt apply criteria not unlike those of early investors in nuclear power, which itself was once an alternative energy source of great promise that would someday make electricity "too cheap to meter." Indeed, Rubin's investors might have been precisely those who wanted to move their money out of nuclear power when it met bad financial times.

Capital employs engineers to design windmills as well as to design nuclear reactors. There are differences in detail, but not in approach. Engineers can and did switch from nuclear to solar energy research quicker and with less explanation than it takes a Yippie to gravitate from the balcony of the New York Stock Exchange to its floor to retrieve the dollars he once threw there in protest. But engineers change quietly.

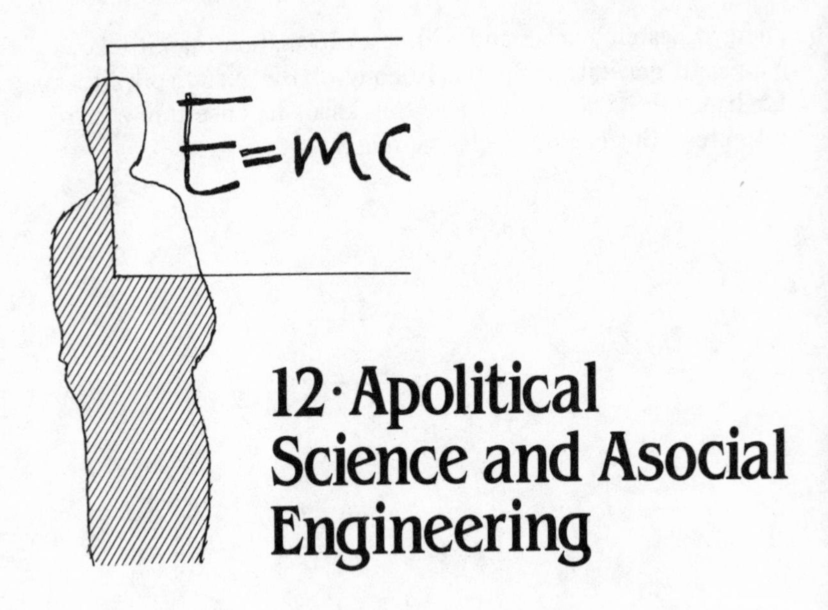

12 · Apolitical Science and Asocial Engineering

From the point of view of accuracy and intellectual honesty the more men of engineering background who become public officials, the better for representative government.

—Herbert Hoover

Science and technology were hot topics as the decade of the 1980s began. The American Association for the Advancement of Science sensed the growing public interest and launched *Science 80,* a magazine that was to change its name annually as if to emphasize the rapid changes that were anticipated in the years ahead. The magazine's success has been phenomenal, and many other new science and technology magazines for the intelligent layman began to vie for the millions of untapped subscribers who were believed to have an insatiable need for information in words and pictures about particle physics, microelectronics, genetic engineering, and other revolutions in the way of life on our planet. One would think a public so eager to know about science and engineering would not settle for sensation without representation, but apparently it does as it always has.

No pure scientist has ever occupied the White House, and the first presidential campaign of this decade certainly did not change that fact. Even before the Iowa caucuses and the New Hampshire primary election it was clear that the list of major contenders would have a predictable composition by vocation: six out of ten hopefuls were lawyers, two were businessmen, one was an actor, and one was a history professor. None of the presidential aspirants listed his occupation as scientist or engineer, both unlikely backgrounds for politicians.

While political pundits did refer to the incumbent's engineering background, President Carter himself never aspired to a technical career. He did attend Georgia Tech, but only to take the mathematics courses he needed to enter the U.S. Naval Academy, from which he received his B.S. in 1946. At that time the driving ambition of James Earl Carter, Jr., was to become chief of naval operations, and his involvement with the nuclear submarine program was just another means to that end. When his father died, the young Carter abandoned his naval career, and he returned to Plains not to practice engineering but to become a farmer-businessman, as he listed his occupation for the record.

The only U.S. president to practice engineering was Herbert Hoover, who worked his way through college and received his bachelor's degree in 1895 with Stanford University's first graduating class. Hoover advanced his career in Australia and China and soon acquired a worldwide reputation as a manager and mining engineer. His series of engineering lectures was published in 1909 as *Principles of Mining,* but the book is certainly less remembered than the translation from the Latin of Agricola's *De re metallica* on which Hoover and his wife, Lou Henry Hoover, collaborated.

Neither book drew much attention from politicians, however. It was in the administrative genius Hoover demonstrated, by successfully managing World War I relief programs, that politicians saw his presidential timber.

President Hoover viewed government as a science, and he wanted to be known as a social engineer who manipulated forces of production and distribution to serve human welfare. But if government is a science, it is not an exact one, and Hoover's failure to be reelected is commonly attributed to his lack of political experience and instinct. By his own admission, Hoover had little taste for playing politics, and it is ironic that the humanitarian relief administrator did not want to give direct federal aid to individuals (of voting age) when the Depression struck because he feared such aid would weaken their moral fiber.

The example of President Hoover is at best problematic, and if there are to be generalizations about scientists and engineers in the Oval Office, there must first be a larger sample of presidents who have come from real technical backgrounds. But the immediate prospects for that larger sample are dim.

The principal sources of presidential candidates are the rosters of state governors and the rolls of the U.S. Congress, but at the beginning of the decade these had a strikingly similar occupational composition to the rosters of past presidents and of presidential candidates. It is as if there were a natural political law always requiring a majority of lawyers and a paucity of scientists and engineers in high public office. Three of every five past U.S. presidents have been lawyers, and the same proportion occupied the executive mansions of the fifty states in the not atypical year in which an actor playing politician successfully defeated the erstwhile engineer turned peanut farmer. In that campaign only Governor Dixy Lee Ray of Washington, who has a Stanford Ph.D. in zoology, could be considered to have had experience in science comparable to Hoover's in engineering. She taught at the University of Washington for over twenty-five years and her experience as chairperson of the Atomic Energy Commission made her technically outstanding among her peers, but she became Washington's only first-term governor to not gain reelection. Her defeat was attributed in part to her style and personality, especially when dealing with business and nuclear issues.

Nuclear energy lost another vocal champion when Washington's veteran representative Mike McCormack, who had had twenty years' experience as a research scientist at an atomic energy laboratory prior to going to the Capitol, also did not gain reelection. And the composition by occupation of the Ninety-sixth Congress was not unlike the historical record of the presidency. Sixty-five percent of the senators listed their occupations as lawyer, while only two could even remotely be considered scientists. These are the former astronauts Jack Schmitt of New Mexico and John Glenn of Ohio. Senator Schmitt, who holds a Ph.D. from Harvard, listed his occupation not only as astronaut but also as consulting geologist. Senator Glenn gave his second occupation as business executive, a category shared by about twenty percent of U.S. senators.

The House, always with a more provincial constituency and with fewer presidential aspirants, comprised only slightly less than fifty percent lawyers. Yet there were proportionately fewer scientists and engineers in the House than in the Senate at the start of the decade of the microcomputer. The four representatives with technical training are noteworthy: Ritter of Pennsylvania, an engineer with an MIT Sc.D.; Martin of North Carolina, a chemist with a Ph.D. from Princeton; and the freshmen, Evans of Iowa, an engineer (B.S., Iowa State) with Pentagon experience, and Coyne of Pennsylvania, a chemical engineer (B.S., Yale) and chemical company executive (M.B.A., Harvard). But these are anomalies and the number of congressmen with technical backgrounds was equaled by the four professional athletes and four funeral directors who occupied seats in the 435-member House. The overall composition of the Congress, which voted on research-and-development budgets in excess of thirty-five billion dollars and highly technical defense systems involving even more money, was less than one percent scientists and engineers. In sharp contrast, it has been estimated that one of every four top-echelon positions in government service is filled with a person who holds an engineering degree. Why do technically trained people so readily fill the ranks of the generally anonymous bureaucracy while apparently eschewing or failing to be elected to nationally visible leadership roles?

It is facile and spurious to argue that lawyers, being versed in the law per se, should dominate the presidency and Congress, for these branches of government represent the people and not the law. In today's times, when increasingly complex technological issues come constantly before the elected officials in Washington, the electorate should be demanding representatives qualified to understand the technical bases and implications of the legislation before them. But the voters do not seem to be looking for technical candidates to vote for and, in spite of the obvious contributions that scientists and engineers can make to the legislative process, the overwhelming evidence is that relatively few run for major public office, and even fewer succeed in getting or staying elected. No doubt the different innate temperaments of scientists, engineers, and lawyers account for much of the way things have been: technical professionals are stereotypically inarticulate loners, and the image of the lawyer is that of gregarious orator.

Not only does the training of scientists and engineers differ from that of lawyers, but also the professions attract different personalities. The ambitious big men on campus who received bachelor of engineering degrees with me two decades ago often had had enough of impersonal equations and graphs by graduation, and they went on to get their J.D. or M.B.A. degrees. It was generally the taciturn and shy among us who stayed in engineering and worked toward the M.S., Sc.D., and Ph.D. degrees that removed the quietly ambitious among us still further from the worlds of business and politics. And the same seems to hold generally for scientists. Any political drives that scientists and engineers have seem to be satisfied within their professional societies.

The professional schools seem to have often reinforced the stereotypes. Certainly engineering schools in their traditional forms have not made it any easier for their students who might aspire to political office to realize their aspirations. The rhetorical skills prerequisite to a successful political career are as dulled in engineering schools as they are honed in law schools. Not only is the law student more likely to attend lectures delivered by masterful rhetoricians, but he is also, more likely, expected to put his own ideas, if not into plain English, at least into terms emotionally convincing to a layman sitting on a jury. That same layman sits on the jury of the electorate every November.

Technical education, on the other hand, has often focused on the manipulation of exotic and esoteric symbols and equations on a blackboard or projection screen, and the students have been trained to communicate not through the spoken word alone but through the media of graphs, diagrams, and equations—with words used mostly as punctuation and conjunctions. The modern technical lecture, perhaps the paragon of technical communication today, often takes place in a darkened auditorium with the speaker's back to an audience composed of his technical peers. Even if the speaker does look at his audience and not at the slide projected on the screen, his audience does not look at him but at the wall, as if they were watching the shadows in Plato's cave. This is a far cry from the legal paragon in which a famous trial lawyer looks a jury in the eye and puts the pertinent legal theorem into eloquent prose they all can understand.

So if a lawyer is trained to argue his case before a jury of citizens who are predisposed to see him as getting to the bottom of his case, what better preparation for a political campaign? And what an opponent for a scientist or engineer who is used to talking in jargon and acronyms to his peers and whom the voters perceive as either talking above their heads or behind his own back, as he literally does. Either way, the citizen is left feeling he is in the dark, as he literally would be should he attend a technical lecture.

I can remember not one instance during nine years of my own engineering education when I was asked to imagine myself explaining something technical to the layman. At the time, the term *science* or *engineering popularizer,* more often than not, carried a pejorative connotation among professionals, and the magazines *Popular Science* and *Popular Mechanics* were dismissed as if they were pulp fiction. Now, twenty years later, there is recognized among engineering and scientific as well as liberal arts educators a need to take engineering and science to the technically literate public and to educate the technically illiterate. Perhaps this enlightened attitude will produce technically savvy political candidates—and voters—in due time.

If the potential engineer- or scientist-politician has had few role models, the electorate has had even fewer images of the technologist as leader or hero. Certainly technical education has no popular counterpart to "The Paper Chase," the television series that sympathetically dramatized and glorified the arduous path through law school. In that series, the demanding law professor inculcated the high principles of legal practice into a receptive class of hard-working students whose interests, by the end of each episode, if not at the beginning, were moral and altruistic.

Television, the modern myth machine, has projected lawyers, like Perry Mason, almost into folk heroes. But engineers, while they are responsible for the behind-the-scenes development and implementation of the medium itself, are as scarce on the home screen as they are in elected public office. Although individual scientists have become popular television celebrities as guides through such ambitious adventures as the ascent of man or the exploration of the cosmos, these are more the impractical pursuits of poets and not of politicians, for they are not generally perceived as relevant to the hard realities of budgets and bargains struck on Capitol Hill. Except for medical doctors, who are not grouped with scientists in the common viewer's mind, television scientists are either obsessed with science itself or mad at the world. Thus when the scientist is placed in the human context of politician, his television image does not win him votes.

The result of all this is that, although the broader society is getting more and more technological, it is far from a technocracy. The statistics substantiate that this was true at the beginning of the decade, and the traditional training of scientists and engineers has not encouraged them to take their cases to the public in realistic terms when they are inclined to seek public office. As the Ninety-ninth Congress convened at mid-decade, the lawyers among House freshmen outnumbered the sole engineer thirteen to one. And the whole Congress of 535 senators and representatives included almost 250—or nearly half—lawyers, but there were still only about a dozen or so, depending on how one counts, engineers and scientists. While the business of Congress continues to become more and more technological, the Congress itself certainly does not. Only the changing nature and image of technical education holds any promise of changing that in the foreseeable future.

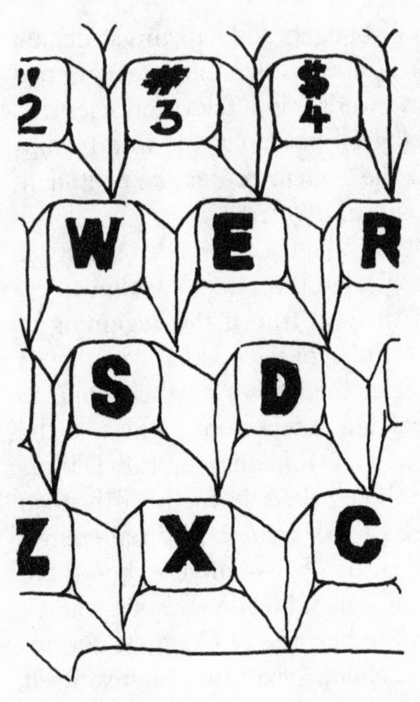

13 · Numeracy and Literacy: The Two Cultures and the Computer Revolution

I

For a long time engineering students were stereotyped as social misfits who were ignorant or insensitive to the traditional culture of the humanities. Today, however, the humanities student is being challenged because he or she is technologically illiterate, or not numerate. While no engineering student can graduate from an accredited school without taking approximately one-eighth of his or her courses in the humanities and social sciences, many a history or English major earns a degree without nearly so much mathematics or science, let alone engineering. Thus a curious change of perception has occurred as to who is and who is not liberally educated in an increasingly technological age.

One thing that seems to have triggered the change of perception is the personal computer. This object of the highest technology is held out as the sine qua non of the twenty-first century, and no one—engineer, scientist, or humanist—wants to be left out of the New Year's Eve party in 1999, or, more precisely, in the year 2000. Scientists and engineers are perceived as knowing all about computers, which interminable advertisements tell us are now not only for calculating numbers but also for processing words. And since words have for so long been the domain of the humanists, the humanists have taken note. This is a propitious development, for the computer is indeed an excellent machine to provide a bridge between the two cultures. Whether it be used for numbers or words, the personal computer tempts its user to cross over to the other culture. It is like using a dictionary: you go to it for one purpose, but you browse about and cannot help but learn something else. So the engineer and scientist who have gone to the computer for their calculations have found that it is a wonderfully patient writing companion with whom they can explore the English language. And the humanist who has gone to the computer as a word processor has found it unavoidable to learn a bit about the science and technology of the device, lest he lose the words entrusted to its memory. Thus one of the unforeseen contributions of the personal computer may be to get the two cultures to come to this bridge from their separate sides and to meet at the middle. Such a meeting has been a long time coming, and in the meantime engineering and technology have emerged out of the shadow of science, with which they have so often been confused.

II

The public perception of science (and technology then as applied science) was radically altered by the development of the atomic bomb, which showed not only how powerful could be the fruits of science but also how clandestinely scientists (and nameless engineers) could work toward revolutionary ends. If secret cities, laboratories, and factories could be built, if basic physical theories could be developed into pieces of hardware that could obliterate tens of thousands in a single flash, if millions upon millions of dollars could be spent, and if tons upon tons of silver from the U.S. Treasury could be diverted to be used in giant magnets designed to separate isotopes of uranium in an effort dubbed the Manhattan Project, then what else might scientists and a government do without consulting the governed?

By the 1950s there was an international paranoia that abuses tolerated in one government during times of war would be extended by others to times of apparent peace. Even before Sputnik was launched there was considerable concern in the West that the Soviets were training more applied scientists. The federal research and development budget was growing by unprecedented amounts, and science was being accepted more and more not only as a worthwhile endeavor but also as a seedbed from which would spring the fruits known as technology. Although scientists always desired free rein and argued for the purity of the research they conducted without regard for its consequences, the reality of the atomic bomb, and its unnamed and unimagined but imaginable successors, was lost on no one. The war had made urgent an issue brewing at least since the earliest days of the Industrial Revolution: how to integrate the newer scientific-technological knowledge with the traditional humanistic culture?

James B. Conant, president of Harvard at the time, excused himself from explicitly discussing the bomb in a series of lectures he delivered in 1946. Instead of worrying for the moment about that one manifestation of scientific and technological development, he dealt with the more general problem of scientific literacy. He summarized his argument as follows:

> . . . we need a widespread understanding of science in this country, for only thus can science be assimilated into our secular cultural pattern. When that has been achieved, we shall be one step nearer

> the goal which we now desire so earnestly, a unified, coherent culture suitable for our American democracy in this new age of machines and experts.

In spite of his mentioning only science, Conant's reference to machines suggests that he also had *technological* literacy in mind, though in the spirit of the times of "applied" science, scientific literacy must have been assumed sufficient to complete an education. This view was to prevail for some time.

In a 1955 speech before the British Association for the Advancement of Science, Jacob Bronowski looked ahead to the mid-1980s, saying, "It is certain that the educated man of 1984 will speak the language of science." Bronowski went on to plead for widespread scientific education as the antidote for a society in which scientists rule the scientifically illiterate. Science, Bronowski argued, is a part of culture and thus should be a part of the education of those who might be in power, just as scientists should have as part of their education the rudiments of literate, humanistic culture. The problem of balancing the sciences and the humanities in the ideal curriculum was reawakened.

When Bronowski called for science as part of the education of a cultured person, he was merely raising anew the debate that Thomas Huxley and Matthew Arnold engaged in in the nineteenth century. Two centuries after the beginnings of the Industrial Revolution, it was time for the issue to arise again, especially since scientists were not only claiming knowledge of the innermost workings of the mind and of the atom but also were claiming larger and larger federal appropriations. Artists and humanists had certainly never known that kind of public financial support, even when they worked on their federally funded New Deal projects.

While Bronowski called for science to be accepted as a legitimate part of culture, he largely spoke to scientists and told them what they were predisposed to agree with, if not fundamentally at least territorially. He called for the educational process to include such specifics as statistical methods, statistical mechanics, atomic physics and chemistry, the scientific method, and even a small piece of personal scientific research for every boy and girl. But Bronowski's speech, even though entitled "The Educated Man in 1984," made little impact outside the scientific establishment. Although Bronowski distinguished the newer scientific culture from the more traditional humanistic one, his speech was largely ideological and abstract. The speech's memorable anecdote concerned Winston Churchill's directive in 1945 to launch the British effort to make an atomic bomb. According to Bronowski, Churchill confessed to being personally quite content with existing conventional explosives but could not ignore the technological counsel that the new weapons would make the conventional virtually obsolete. Bronowski's point was to emphasize the importance of technologically informed decisions, which are possible in a democracy, he argued, but not in a dictatorship where the contentment of the leader could jeopardize the welfare of the citizens.

About a year after Bronowski's appeal for science as part of culture, a short essay by C. P. Snow appeared in the 1956 Autumn Books Supplement of *The New Statesman and Nation.* This essay began on much the same premise as Bronowski's, that science was somehow just not quite considered a part of culture, but while Bronowski emphasized sound educational reform, Snow related gossipy anecdotes and dropped names freely. Furthermore, he did not so much spell out concrete proposals for incorporating science into culture as he pointed out and lamented the differences between the scientific and the literary cultures. Snow entitled his essay "The Two Cultures."

There was remarkably little reaction to Snow's argument in the correspondence department of *The New Statesman.* There were a few letters, but nowhere near the volume of mail elicited by a report, published only a month before Snow's essay, calling for more students to study science instead of the humanities. The study, entitled "New Minds for the New World," raised the specter of the Soviet threat and documented the paucity of pure and applied scientists being turned out by Britain as compared with the Soviet Union. The report called for more young people to study science as it confessed: "Our present culture is, of course, unscientific. It is also in part anti-scientific."

Though Jacob Bronowski and C. P. Snow were lobbying on both sides of the Atlantic for the intellectual legitimization of science, humanists apparently did not feel particularly threatened in their dominant cultural position. F. R. Leavis, the prominent Cambridge literary critic, was being criticized for his lecture, "Literature in My Time," in which he declared that, apart from D. H. Lawrence, there was *no* literature in his time. This prompted the novelist J. B. Priestley to write some scathing "Thoughts on Dr. Leavis" for the pages of the same volume of *The New Statesman* that contained Snow's claim to cultural status for the science of the time. Unlike the few quiet letters responding to the idea of "two cultures," however, the Priestley–Leavis debate generated volumes of mail that appeared for months in the letters column of the magazine.

Yet, not until Snow expanded his slight essay into the Rede Lecture, published as *The Two Cultures and the Scientific Revolution* in 1959, did anyone really seem to take notice of the idea Bronowski articulated in 1955 and Snow christened in 1956. Leavis had ignored Snow's earlier essay in *The New Statesman* but took exception to the attention given to a new scientific culture, especially if it were at the expense of the traditional literary and historical culture. He delivered a terribly ad hominem lecture, entitled "Two Cultures? The Significance of C. P. Snow," which was subsequently published in *The Spectator.* Leavis found Snow "intellectually as undistinguished as it is possible to be," and "in fact portentously ignorant." *The Two Cultures,* Leavis said, "exhibits an utter lack of intellectual distinction and an embarrassing vulgarity of style."

What had incited Leavis in the few years between Snow's modest essay and the Rede Lecture? Well, when the Russians launched Sputnik, all the pedagogical warnings about the Soviet edge in technologists and all the theoretical musings about science being a new part of culture were suddenly matters of political and ideological life and death. The artificial moon beeping down at the polarized Earth made the issue of who was and who was not a great contemporary novelist seem among the most trivial of pursuits.

III

C. P. Snow's "two cultures" are sometimes envisioned as islands populated by those conversant in either Shakespeare or thermodynamics but not in both. Scientists, sharing a common "method," were seen by Snow to have "the future in their bones"; the humanists, represented principally by the literary intellectuals, were "natural Luddites" who had "never tried, wanted, or been able to understand the industrial revolution, much less accept it." And this distinction was crucial to Snow's message, for he saw science and technology as not only the principal ingredients of twentieth-century material wealth but also the necessary ingredients in the development of the Third World. Asian and African countries were ready for their own industrial revolutions, and Snow described as "suicidal and technologically illiterate" those products of the old humanistic culture who thought the underdeveloped countries should be content to creep up to the standards of the industrialized countries in a matter of centuries.

The issue that Snow delineated was that of the informed use and dissemination of technology, and he echoed Bronowski's call for educational reform. Yet even as Snow was delivering his lecture on the two cultures, engineering students were required to fill the equivalent of one full semester of their four-year curriculum with social science and humanities electives. And since Snow's Rede Lecture we have had: the rise and fall of federal research-and-development budgets; the rise and fall and rise again of engineering enrollments; the infusion of foundation grants to colleges and universities to infuse even more humanities and social sciences into science and engineering education. These grants have spawned numerous programs, generically termed Science, Technology, and Society programs, which have formalized and institutionalized interdisciplinary activities of faculty and students. If not their intent, at least their effect has been that of humanizing scientists and engineers rather than informing humanists about scientific method and culture. Only very recently has there been a strenuous effort to introduce science and technology into the traditional liberal arts curriculum.

As C. P. Snow even acknowledged in his Rede Lecture, to divide the educated world into two distinct cultures is at best arbitrary. Social scientists presented a special difficulty to Snow's dichotomy, and in his "second look" he admitted that a "third culture" was on the way. Snow believed this third culture to comprise a body of intellectual opinion he saw then emerging—from such fields as sociology, political science, economics, and psychology—"concerned with how human beings are living or have lived." But some other nontraditional disciplines, which could presumably also provide a bridge between Snow's two cultures, seem sometimes to be without firm anchorages in either one. And signs of weakness do not bode well for the success of bridges.

A recently published book, *Global Stakes: The Future of High Technology in America,* states that its purpose is "to pull together in one place the facts and figures, arguments and opinions, that affect the future of high technology in America. . . ." The book's title page identifies three authors writing "with" a fourth, and a back page "about the authors" identifies them as international consultants on policy. What makes the book relevant to the present discussion of C. P. Snow's pivotal lecture is *Global Stakes*'s summary of the two cultures debate in a chapter entitled "The

New Education." Not only is Snow identified as an "English philosopher," but also his allusion to Shakespeare and thermodynamics is related with a reference to "Newton's Second Law of Thermodynamics." Now even those literary intellectuals who disagree with Leavis about whether Snow was a novelist would certainly not dub him a "philosopher," and no scientist has ever confused the Second Law of Thermodynamics with Newton's Second Law of Motion. This is an amazing pair of gaffes to have escaped four authors and, one should think, at least one editor suspended somewhere between cultures of which they seem possibly never to have been a part.

For all their presumed faults, the allegedly incompatible spokesmen for each side of the familiar two cultures are generally, if nothing else, sticklers for detail. That is to the good, for when the details get fuzzy, the thinking gets fuzzy. Indeed, it might be argued that the world, the Third World included, might better be served by a pair of separate but equal cultures of careful specialists than by a gaggle of generalists who cannot go beyond a sloppy explication of the issues. When the times demand it, the humanists can sit down with the scientists, whether or not there are members of a third culture present, and meet on the common ground of appreciation for discipline. For in the end it is discipline, the rigorous attention to the detail of one's work, that distinguishes the scholar from the dilettante. That is not to say that one cannot reach beyond his discipline, only that one must do it with the same care and rigor that he pursues his first discipline.

IV

When the Alfred P. Sloan Foundation published its recent occasional paper on "The New Liberal Arts," it reopened the question of acculturation. Some twenty-five years after Jacob Bronowski and C. P. Snow pleaded for science and technology as parts of culture, a foundation was finally willing to make a financial endorsement of the idea. It is time, argued Sloan spokesman Stephen White, not only for the engineer and scientist but also for the general liberal arts student to understand the computer, applied mathematics, and engineering, since the future is clearly and inextricably involved with such subjects. Exactly how to implement such a radical change in the traditional liberal arts curriculum is another matter, however, and colleges and universities have been invited to search their souls and budgets over the matter in proposals to the Sloan Foundation.

While applied mathematics and engineering are not new and may remain esoteric and foreign disciplines to the liberal arts student (and faculty) for some time to come, computers have quickly become ubiquitous and thus hard to ignore. And the benign presence of the computer in homes, schools, and businesses everywhere may do more to bring about the integration of technology into the mainstream of culture than all Bronowski and Snow's pleas, as well as all the well-intended efforts of government agencies and private philanthropy. In particular the perfectly named personal computer, the instrument that embraces at once the human value of individuality and the highest technology of the late twentieth century, has the capability to bridge the gap between the two cultures—and with plenty of clearance for the great commerce of all the other cultures to continue unobstructed.

The personal computer displaces at the same time the typewriter of the humanist and the calculator of the engineer, yet it can provide a common ground for mutual understanding. Whether used as a word processor or a number cruncher, the microcomputer is a robot shuttle diplomat. It moves freely back and forth between the cultures, alternately speaking the language of the writers of prose and of equations. And in its keyboard the personal computer unites the objects of literacy and numeracy in a way heretofore unrealized.

The classic typewriter keyboard, with its QWERTY arrangement of keys (said to be laid out deliberately to impede their too rapid depression so that the original mechanical contrivance would not be forever getting tongue-tied), was never hospitable to the ten digits, which always seemed out of reach and never quite in the repertoire of the touch typist. Though they appeared to be in strict order, so unlike the jumble of letters, the numbers were struck with the eye, and not with the finger alone. Equations were a nightmare to type, and many an old keyboard did not even have an equals sign. More than one model expected the user to employ a lower case *l* for the numeral *1*, while the capital *O* and the cipher *0*, though on different keys, were indistinguishable. Such inconsistencies were adapted to in the same way cotton ribbons were tolerated.

The advent of the computer brought with it the necessity at once to combine and separate more the alphabet from the numbers. Early computers did not read marks on paper, but holes in tapes or cards. Each letter and number had its own combination of holes and not holes, of zeroes and ones, of ons and offs, for in the final analysis that is all the electronic brain is capable of distinguishing—the openness or closedness of switches, the *yeses* or *noes* of precedence. The typewriter's little *l* could no longer be ambiguous, and the big O and the zero had to be easily told apart. The numeral *1* was added, and the numeral *0* was distinguished from the letter O the way a corporal is separated from a private—by a stripe.

If the typewriter favored the humanist, the modern pocket calculator favors the numerically inclined. Its keypad puts the ten digits within easy reach of one set of fingers, and in a tight numerical order. As secretaries learn the typewriter keyboard, so human calculators learn this keypad:

7 8 9
4 5 6
1 2 3

with some variations in the placing of the zero and the decimal point on the bottom line. This is a very sensible arrangement of the keys, with the digits ascending from the zero in order. One's finger must reach higher for a higher number.

The telephone dial, like the typewriter keyboard, was another technological artifact that confused the alphabetic and numeric symbols. It was not the letter *O* but the digit zero that stood for "Operator." Furthermore, the rotary dial was a classic example of innumeracy that, like the QWERTY keyboard, puts *0* after *9* instead of before *1*, where it properly belongs. And a curious dilemma appears to have developed with the touch-tone telephone number pad. Meant incidentally to allow us to enter our charge and bank account numbers, so we might pay and save by phone, the touch-tone number layout retained the vestiges of the alphabetic telephone exchanges in the letters associated with the number holes on the old rotary dial. To arrange the telephone buttons in the same numerical order as the calculator would have introduced a very peculiar arrangement for the letters of the alphabet. If the groups of three letters associated with most of the number keys are ever dropped from the telephone buttons the way exchange names have by and large already been dropped, the numerate society may make a trivia question out of the numerology of telephone buttons as they are arranged today:

1	2	3
4	5	6
7	8	9
*	0	#

This disagreement in the arrangement of calculator and telephone number buttons is as unfortunate as if a major typewriter manufacturer departed from the QWERTY keyboard arrangement. Only certain keyboards intended for the use of children, who are expected to hunt and peck at the letters and numbers, are arranged in strictly linear alphabetic and numeric order. What large computer manufacturers have adopted is a dual keyboard. Not only is the conventional QWERTY keyboard there for the humanist, but also the number pad with the digits in calculator order is given a separate place off to the right of the "typewriter" section, allowing masses of numerical data to be entered and calculations to be made by the touch method. And just as humanists and technologists have in the past ignored the other culture and its tools, so human word processors and human number processors may now ignore the part of the keyboard that is not their immediate concern.

The computer and its keyboard thus provide an excellent metaphor for the two-cultures problem, and indeed for the problem of knowledge and disciplines generally. By containing the twenty-six letters of the alphabet and the ten digits, the keyboard presents to the user the potential to express and explore any literate or numerate idea. At the same time, the redundancy and duality of the keyboard enables one to ignore whole segments of the machinery of thought and expression, and it is possible for one to completely reject the access he has to an entire half of the electronic brain.

Yet the computer and its keyboard provide not only a metaphor but also a means for twentieth-century man to overcome the two-cultures problem in ways that Bronowski in his eloquence and Snow with his anecdotes did not foresee. While the computer may be considered expensive by some standards (notably, the cents sign is perhaps the only symbol from the old typewriter not on the keyboard of the most popular personal computers), it is in fact already within the purchasing grasp of the great middle class and promises to become less and less costly in the same evolutionary way that the pocket calculator has. Thus we can realistically look ahead to a time well before the turn of the century when the computer may be as familiar an appliance in the home as the television set.

V

Because of its essential duality, the home computer can dissolve the distinctions between the two cultures. The humanist who wants to word-process may choose to ignore the number keys, but he cannot ignore certain technical details: the capacity of his machine's memory, what kind of floppy disks it uses, and the technical procedures to save and retrieve his words. If he wants to print out his manuscript, he must find a compatible printer, and if he wants to do things right he may even read the manuals and user's guides (poorly written and incomprehensible as they may seem to be). By involving himself with this technology, for it is that if nothing else, the humanist begins to appreciate the other culture with neither awe nor disdain. He begins to understand that technology is *human*, that is, it has limitations.

No one can use a computer without coming to the realization that the human race has nothing to worry about. While the capabilities of the computer may seem limitless at first flash, the writer who has to type in more than a short manuscript soon realizes how limited many a machine's memory really is. While it may be long, it is not deep. A floppy disk cannot hold a novel, nor can the computer plot it. What it can do is manipulate the words and paragraphs in a terribly obedient way, so that the writer need not be intimidated by the task of retyping the whole manuscript, even the long parts that he does not want to change. The computer can be called upon to check spelling, but the writer soon learns that electronic spelling checks do not catch the mistyped or misused *principal* for *principle*, for both agree with the dictionary they are checked against. The non-technologist who uses a computer soon learns how silly are the fears that this device can ever replace man.

The technologist, on the other hand, who for so long might have eschewed typewriters as the tools of workers in words and not in equations, cannot help but be drawn to the QWERTY hardware and the wordy software that enable him to hone his thoughts into sentences that can be considered humanistic. The tireless video screen will show draft after draft of his latest paper without calling for a halt the way secretaries are wont to do. With the word processor loaded in his computer, the engineer or scientist can seek the elegance in his prose that he seeks in his equations. And with practice at the computer terminal, he can get it.

When and if the technologist masters the English language, he will come to appreciate that the humanist's traditional tool demands as much precision in its proper use as do the tools of science and technology. For all the narrow-minded conventional wisdom, there is more latitude in employing the mathematical language of calculus or a computer language such as BASIC than is commonly allowed. True, there are rules that must not be violated if one is to achieve valid results, but so there are rules of grammar that one must not violate if his humanistic argument employing the English language is to be taken seriously. If the computer will not accept a command containing a syntactical error, neither will the referees of a scholarly journal be inclined to accept a paper in history, say, if it is filled with spelling and grammatical errors. Anyone who admits the importance, or at least the desirability, of numeracy should recognize the equal importance, or desirability, of literacy—and vice versa. Discipline demands discipline.

Technology is here to stay. Its elements are as much a part of our culture as those of the traditional humanistic disciplines. The modes of thought of engineers and scientists, the languages of the physical, biological, and computer sciences, and the muscular and mental disposition required to operate a personal computer as naturally as one operates a motor vehicle must become essential to the educated man or woman in the closing years of the twentieth century. Without these we cannot deal confidently with such issues as nuclear power and nuclear weapons, genetic engineering, acid rain, telecommunications, energy, or even budget deficits and proposals to eliminate them.

Jacob Bronowski's expectations for the educated person of the mid-1980s have not yet been realized, but at least now they are more than one person's hopes. The computer, though not even mentioned by either Bronowski or Snow a quarter century ago, may indeed prove to be the mechanism by which we will both broaden the concept of tradition and bridge the gulf between the traditionally antipathetic cultures. The new intellectual must recognize that keeping up with advances in science and technology is imperative for those who wish to march in the avant-garde of tomorrow.

14 · A Diary, Exhibiting Some Reasons Why There Is and Some Reasons Why There Should Not Be an Engineering Faculty Shortage

Monday, September 26

7:00 A.M. Awake to alarm. Make sure Karen and Stephen are awake.

7:05 In shower, think about week ahead. Review today's schedule while shaving. Identify time to prepare tonight's lecture.

7:25 Look through clean shirts with week's schedule in mind. Choose most comfortable one.

7:30 Breakfast of juice, cereal, and coffee. Look through morning paper.

7:50 Suggest to Catherine that she and children have pizza for dinner. I will eat with group sponsoring this evening's lecture.

8:00 Walk to university. Rehearse lecture along way.

8:25 Enter School of Engineering and see first posters advertising my lecture, "The Crystal Palace of 1851." Note posters make no mention that lecture is at University of North Carolina and not at Duke. Check mailbox. Find nothing but a new roster of graduate students. Notice omissions. Pencil in additions and leave copy for Mrs. H. to make changes.

8:30 Go to Xerox machine to copy manuscript promised to be mailed today. Find copier, broken last Friday, still not working.

8:35 Enter office to desk covered with graduate student's calculations and computer results. Recheck derivations of equations, logic of computer program, and plots of results. Puzzling behavior of cracked cantilever beam remains unresolved.

9:40 Interrupted by phone call canceling meeting scheduled for later this morning. Discuss subject of meeting, the promotion of an assistant professor, briefly on phone. Agree to reschedule meeting when dossier complete.

9:55 Check mail room. Find Saturday's U.S. mail has been distributed with this morning's campus mail.

10:00 Open mail: Letter from university press saying they do not specialize in books on engineering education and will return my manuscript under separate cover; letter of acceptance of technical article by *Res Mechanica;* announcement and call for papers for Sixteenth International Conference on Theoretical and Applied Mechanics to be held near Copenhagen next August; memo announcing agenda of Thursday's meeting of Engineering Faculty Council; revised roster of graduate students; inquiries about graduate program from India, Taiwan, Canada, and New Jersey.

10:10 Cull inquiries for Jo Ann to prepare standard replies.

10:15 Return to problem of cracked beam. Discover student has derived equations for more special case than intended. Also find apparent error in transcribing formula from a handbook to computer program. On further checking find handbook contains typographical error.

10:50 Interrupted by colleague who objects to graduate student roster identifying some master's students as doctoral students before they have passed final examination for M.S. degree. Agree to do something about the oversight.

11:05 Track down Mrs. H. and find rosters already distributed not only to faculty but also to all graduate students. Say I will figure out later how to fix.

11:10 Return to cracked beam problem. Annotate computer program in preparation for conference with graduate student.

11:15 John Fast, an undergraduate to whom I serve as adviser, stops in to chat between classes. He asks if I have seen latest issue of *Forbes* with article on "science" of hitting baseballs.

11:35 Begin to arrange transparencies for evening's talk. Decide to start by discussing nineteenth-century iron bridge failures and then offer Crystal Palace as example of Victorian structural success, thus arguing that bridge failures were not due solely to engineering ignorance.

11:50 Interrupted by colleague sorry he cannot come to talk this evening.

12:25 P.M. Check mail room. Nothing but some drawings draftsman has revised for third time. Pick up today's *Chronicle*, the student newspaper.

12:30 Eat lunch of grapes and cheese while reading about weekend meeting of Board of Trustees. New ten-million-dollar dormitory project employing fast-track construction technique approved in principle. Make mental note to clip articles and start file on possible case study of success or failure of fast-track method.

12:50 Return to transparencies of Crystal Palace and note place where construction time (seventeen weeks from erection of first column) can be mentioned. Notice that there is no transparency of New York Convention Center, whose design is based on Crystal Palace, but whose fast-track construction schedule has slipped a year and whose cost has almost doubled. Decide there is no time to hunt down picture.

1:15 Check mailbox and find lots of stuff. Remember I was to mail manuscript to editor. Find Xerox machine fixed; copy manuscript on effects of microelectronics revolution on family and mail to editor who suggested I write the piece.

1:25 Open mail: Letter congratulating me on Crystal Palace article in *Technology Review;* copy of journal containing excerpt of another article of mine, along with letter from permissions editor of *Technology Review* regretting that neither author's affiliation nor copyright were properly noted on the reprint; several announcements of new books; copy for upcoming edition of *Who's Who in the South and Southwest;* collection of inquiries about graduate study.

1:40 A graduate student comes by to say he would like to attend my lecture but posters only went up this morning and he had already made other plans for the evening.

1:45 Look at material for tomorrow's classes. Begin to prepare the morning lecture.

2:35 John Fast stops by to ask questions about impact loading. He draws on blackboard and describes problem he is trying to solve: a lock struck by a sledgehammer.

3:10 Afternoon campus mail has brought notice from benefits and records division that Blue Cross/Blue Shield premiums will be raised effective November 1, thus reducing October paychecks. Graduate School has forwarded inquiries about graduate programs in civil and environmental engineering.

3:15 Drop off pile of inquiries for Jo Ann to process. Pick up for signing couple dozen letters to prospective graduate students in response to inquiries already processed.

3:25 Go over transparencies for Crystal Palace talk.

3:50 Interrupted by phone call from colleague who wonders why he is listed on roster as adviser to a graduate student who is not doing any research for him. Say I will talk to student and try to clarify matter.

4:10 Try once again to go through transparencies without interruption. Succeed in reviewing an hour talk in fifteen minutes.

4:35 Leave office for home.

5:10 Wash up. Talk to Catherine and children about the day.

5:20 Leave home for Chapel Hill to meet hosts for dinner.

5:40 Park car and meet young man who had read my article in *Technology Review* and asked me to talk about Crystal Palace before the Dialectic and Philanthropic Literary Societies. He introduces young lady who is another officer, and we wait on Franklin Street for others to arrive.

5:50 Go down alley to enter popular student restaurant in basement that seems to extend under Franklin Street. Discuss logistics of getting an extension cord. Talk is scheduled for Dialectic Hall in a building called New East, which dates from the time of the Crystal Palace and has few electrical outlets.

6:00 Remaining officers of the combined societies arrive and we order. Ask for suggestions. Host recommends flank steak described on an irreverent menu as "The Sizzler." Since only heavily sweetened iced tea is available, order a beer to drink. We are treated to a monologue history of the Di's and the Phi's by the senior member of the group.

7:00 Leave to get extension cord, which is in a sculptor's studio about seven miles south of Chapel Hill. Wonder if we will get back in time for 7:30 talk. Host assures me we will and that I must meet the sculptor and see his studio.

7:15 Arrive at studio but sculptor is nowhere to be found. Host locates hidden key and leads me through old schoolhouse to main studio. Extension cords are strung everywhere and host selects one with least friction tape. He points out busts of Thomas Wolfe, Jesus Christ, Sam Ervin, and other state heroes that sculptor is working on.

7:31 Arrive at Dialectic Hall. Rearrange furniture to cast shadows of the Crystal Palace among portraits of famous Di's and Phi's of the past hanging around the curved wall of the beautiful third-floor room. Use physics textbook to prop up overhead projector.

7:40 Meeting begins with calling of roll of senators and reading of minutes of last meeting. See and hear this is no shy group of students. Normally Di's and Phi's, who come from opposite sides of the state, sit on opposite sides of the aisle in this miniature senate chamber. Tonight they crowd together on one side to see the curved wall squarely.

7:50 Introduced by officer who did not get my history because we were discussing society's at dinner. She reads poster and gives me the floor.

8:40 Finish Crystal Palace talk. Ask for questions. There are plenty from the room of debaters, and they are all good. Afterward faculty adviser introduces himself as lawyer interested in construction techniques used to build the antebellum state capitol.

9:05 Host gives me supply of leftover posters and leads me

through a maze of campus buildings until I get bearings on Franklin Street.

9:35 Stop to pick up six-pack of beer on way home.

9:50 Arrive home as Catherine is retiring. Children are already asleep.

10:00 Read last Friday and Saturday's *New York Times,* which always come in Monday's mail. Clip article on investigation into collapse last summer of Mianus Bridge in Connecticut.

11:45 Take a beer to study and look through notes on structural failures. Pick up working on organizing notes—spanning several years and scattered through several files—into proposal for book I would like to write next summer.

12:45 A.M. Notes are not making much sense after long day. Pack up briefcase for another day.

1:05 After checking children's alarm clocks and my own, go to bed and fall asleep immediately.

Tuesday, September 27

7:05 A.M. Wake to sound of front door opening. Someone is getting morning paper.

7:10 Think about day's classes during shower and while shaving. Continuum mechanics lecture on stress tensor. Elasticity lecture on curved bar.

7:25 Catherine calls through door something about society column.

7:30 Look for shirt and tie combination students have not yet seen this semester. Since it is chilly this morning decide to wear wool tie and tweed jacket.

7:35 Find Stephen in kitchen reading my name out loud from newspaper. Society column has report of last Wednesday's poetry reading and reception, and Catherine and I are described as "both dark-haired and taller than average." Recall what a long day that was: reception hosted by new dean and his wife kept us on campus till almost 7:00 P.M. Pizza children had ordered for delivery by that time had still not arrived when we got home. Delivery boy called at 7:10 saying he could not find our house. With directions he came at 7:15. We left for Paul Zimmer's reading just before 8:00 and got home from reception hosted by the Fergusons about 11:00.

8:10 Leave for campus and think about how to present stress tensor to thirty-two juniors and seniors who are not well versed in index notation.

8:35 Enter School of Engineering and notice most Crystal Palace posters already taken down. Empty contents of mailbox.

8:40 Read inquiries about graduate program. Find note from Jo Ann tallying inquiries so far this month: thirty-nine from American students and eighty-nine from abroad.

8:50 Check drawings for paper on stability of crack and find some legends omitted and some ticks missing from vertical axis. Note additions for draftsman.

9:05 Check calendar for date to schedule exam in continuum mechanics, go over textbook's sign convention for stress tensor and derivation of differential equations of equilibrium, and select homework problems.

9:55 Student from continuum mechanics class stops by to ask if he might do a senior project on developing computer software to help students learn index notation. I tell him it sounds like an excellent idea and that he should discuss it with his adviser.

10:20 Drop off in mail room inquiries about graduate program and revisions for drawings. Empty contents of mailbox.

10:25 Go over copies of journal contents pages distributed by librarian. Note article on normal modes of the modern

English church bell in the latest issue of *Journal of Sound and Vibration* and mark it as one to look up. My manuscript on the cracked bell is being refereed for that same journal, and this article may be relevant to my paper or contain references of which I am unaware.

10:35 Go to class and announce exam two weeks hence. Students ask about some homework problems that have been especially confusing. Try to clarify some points from previous lectures. Do not get as far with the equilibrium equations as planned.

11:50 Class dismissed promptly, for some students have their next class two miles away on East Campus.

11:55 Find mailbox empty. Pick up copy of student paper in lobby.

12:00 M. Eat cheese and plums while reading about *Australia II* and the America's Cup. Read article about new director of International House, Brian Silver, whom I met four weeks ago when one of my graduate students was taken to emergency room after having bicycle accident. He suffered fractured skull and Brian and I tried to get a telephone number for the student's parents in India when it appeared that neurosurgery might have to be performed. Operation was not necessary, but we had a tense late-night vigil in the hospital waiting room.

12:25 P.M. Check mail room and find engineering education manuscript. No other domestic mail.

12:30 Go over elasticity lecture. Look ahead to next topic and plan exam for same week as continuum mechanics.

1:20 Go to engineering library to look up paper on English church bells. Article is interesting but not directly of use. Browse among new books.

1:35 Find responses to graduate program inquiries filling mailbox. Sign letters and leave in Jo Ann's mailbox.

1:45 Go to elasticity class. Ask mending graduate student if he saw article about Brian Silver in paper. Announce exam date. Solve curved bar problem and discuss how it may be used in conjunction with related solutions to solve a whole class of problems via superposition.

3:00 Finish elasticity class.

3:05 Meet with graduate student whose computer results I went over yesterday. Point out difficulties interpreting results and show him calculations. He agrees to make further computer runs even though several theoretical issues remain unresolved.

4:25 Answer Monday's mail. Draft thank-you note to admirer of Crystal Palace article and send carbon copy to editor of *Technology Review,* enclosing poster from Chapel Hill talk. Draft responses to graduate program inquiries that do not fit into standard categories of form letters.

4:40 Put drafts of letters in Jo Ann's mailbox and pick up final drawings for crack stability paper. Ask Mrs. H. to order glossy prints to be sent to journal.

4:50 Catherine comes to drive me home for early dinner. Tonight is curriculum night at Stephen's school.

5:05 Catherine plays telephone message. Editor likes piece on family life in the microelectronic age and wants to run it in the November issue. She agrees to use pseudonym, "Mr. Chips," which Catherine thought up, to add some mystery for local readers. Marvel at one-day mail service and decisive editor.

5:45 Eat dinner and discuss logistics of getting Stephen to soccer practice and back while we are at school. It is arranged that we will take him and coach will drive him home.

7:10 Leave for curriculum night. Between 7:30 and 9:30 follow a condensed version of Stephen's daily class schedule. Meet all his teachers and hear what they will do all year. Finish night understanding why Stephen is tired after his school day.

9:50 Home for the night. Read *New Yorker* before going into study.

10:55 Look through various art books and encyclopedia trying to identify painting of nineteenth-century railroad station that I seem to remember is by Monet. It is excellent illustration of an iron-and-glass structure that I should like to add to Crystal Palace file. (Monet is not to be confirmed until Karen is put on case.)

11:20 Get to desk and back to files on structural failures. Look through essays and articles I have written on the subject over past five years and try to see unifying themes. Try to imagine how all the pieces might fit together into a coherent whole that might interest some publisher.

12:40 A.M. Go to bed without best-seller idea.

Wednesday, September 28

7:00 A.M. Press snooze-alarm button and roll over in bed.

7:05 Reawake to alarm. Check to see if Karen is up.

7:10 Think in shower about cracked beam problem. Realize that there is something basically wrong with whole approach we are taking.

7:35 Dress in same color shirt and tie as yesterday. Realize this combing hair, but leave them on since I will not meet same classes today.

7:40 Get to breakfast table as Stephen is ready to leave for school. Ask how soccer practice went last night and tell about curriculum night. Wish him a good day.

7:45 Eat usual breakfast over morning paper. Say good-bye to Karen as she picks up lunch on way to car pool.

8:00 Catherine returns from taking Stephen to school. Have coffee together and talk for a while. Tell her about meeting this evening and suggest she and the children have pizza for dinner.

8:25 Leave for campus. Think about cracked cantilever and decide to rework derivation for another kind of beam.

8:55 Check mailbox and find some campus mail.

9:00 Put feet up on desk and read mail. Lots of requests for information, application forms, and financial aid forwarded from graduate school. A student from mainland China wants to study structural engineering and is sure she would "bring honour to Duck University." Several students request waiver of application fee since Indian government will only let them send $100 out of country and that has been largely used up in fees for Graduate Record Examinations and Test of English as a Foreign Language. Those students who look promising are referred back to graduate school with recommendation of waiver; others are discouraged from applying. Mail also includes memo from President Sanford announcing Annual Meeting of the Faculty for October 27. New provost will be introduced at meeting. Go to enter date on calendar and find that it is fourth Thursday, date reserved for Duke Press editorial board meetings. Dash off handwritten memo to chairman of board suggesting our October meeting be rescheduled.

9:15 Begin to look at problem of cracked cantilever anew and realize mistake. Since work must be redone from scratch, decide to do calculation for simply supported beam with crack at arbitrary location. Everything goes smoothly until transcendental equation arises. Its solution appears to be straightforward by graphical means, if nothing else. Take a break confident that everything will work out in time.

10:05 Interrupted by telephone. Mrs. H. says one of our graduate students is on line and would like to make appointment to see me this afternoon. Tell her anytime before four o'clock.

10:10 Return to beam problem and begin to attack transcendental equation. It is not as easy as I had thought.

10:35 Look for chairman to discuss problems generated by roster of graduate students. Ask Mrs. H. to leave message that I would like to talk with him when he is free.

10:40 Deliver graduate queries to Jo Ann and pick up for signature several dozen responses to previously delivered queries. Recognize familiar name among letters. He is student we wanted to recruit last year but funds were unavailable then to make an offer. Now there are new funds. Write letter to that effect.

10:55 Get out file of article to appear in *International Journal of Pressure Vessels and Piping.* Its illustrations should be ready by end of week, but letter of acceptance requested an abstract and list of symbols be added to manuscript. Go through paper and write down in alphabetical order each new symbol encountered. Write definitions of symbols and abstract. Draft letter to accompany new material and glossy prints when ready.

11:20 John Fast stops by between classes and talks about science fiction, fall of the Roman Empire, fluid mechanics, and his "broken" foot.

11:30 Look at calendar and realize preregistration for Spring Semester is approaching. A seminar on faculty research must be organized before preregistration so students without thesis topics may be exposed to some possibilities. Identify good time and make note to bring subject up at next faculty meeting.

11:45 Chairman comes in and asks what I wanted. Describe faculty reactions to student roster and suggest we give more thought before issuing another one. Agree that this roster can serve as an excuse for a memo clarifying policy on when a student becomes officially a Ph.D. student and when a faculty member and a student become officially an adviser-advisee pair.

11:55 Begin memo.

12:40 P.M. Check mailbox. Nothing. Pick up a *Chronicle* and copy of *The Missing Link,* "a new student publication hoping to promote student involvement in the intellectual and cultural activities around Duke and Durham."

12:45 Eat cheese and banana over the *Chronicle.* Find letter to editor from student whose name is familiar. She is mourning the "firing" of popular assistant professor whose mechanical design course some students consider a highlight of their curriculum. There is also another letter about the "exemplary teacher" who will not return next year because the university will not grant him tenure. Reminded that specialists in mechanical design spend a lot of time supervising student design projects and competitions and typically do not publish as much as their colleagues in other fields. Imagine how to explain to students that sometimes such teachers perish.

1:10 Go to library for weekly check of latest issue of *Engineering News-Record.* Find letter to editor challenging article suggesting that Severn Bridge hanger wires are failing because it is overloaded. Letter defends original design and questions use of "unproven" analysis over traditional methods of practicing engineers. Xerox letter to throw in possible-book file on desk at home. Make extra copy for graduate student who wrote paper on Severn Bridge and one for colleague who has been asked to study Bosporus Bridge of similar design. Look quickly through last Sunday's *New York Times,* keeping eye out especially for other articles following up on Mianus Bridge collapse or New York Convention Center construction delays. Find nothing. Look in Week in Review section and skim Careers in Education for any interesting open positions. Nothing.

1:40 Collect mail. Chat in mail room with colleague re inquiries from potential graduate students in his area.

1:50 Mail contains confirmation of room reservation from

Hyatt Regency Houston. American Society of Civil Engineers' Structures Congress will be held there in October and I have a paper on large plastic deformations of cracked cantilever beams, which chairman of session, a former colleague at Argonne, invited me to give. Looking forward to seeing him and downtown Houston buildings, including hotel itself, whose windows were broken by recent Hurricane Alicia. Other mail includes ballot from Society for Natural Philosophy, latest issue of *American Scientist,* and inquiries about graduate study.

2:05 Return to memo prompted by troublesome roster. Encounter difficulty in expressing without ambiguity idea that adviser-advisee relationship is one of mutual agreement and that neither party should presume that he is being advised or advising without some explicit mutual agreement having been reached. If memo is written carelessly it may create more problems than it is supposed to solve.

2:35 Deliver draft of memo for typing. Drop off latest batch of queries and pick up dozen or so processed ones for signature.

2:40 Return to transcendental equation. Find it can be solved in principle by iteration on hand calculator. Get out calculator and begin to iterate, but do not get convergence.

3:10 Graduate student who made appointment arrives. He has received notice from bursar claiming that over four hundred dollars in tuition was not paid last December. That was over ten months ago but apparently error in accounting was just discovered. Student understood that all his tuition and fees would be paid by professor's research grant and wants to know what to do. Professor's grant has no more money now. Try to unravel what appears to be an unusually tangled case of student being switched from contract to grant and suggest student see chairman since none of the graduate school funds I administer is involved.

3:30 Telephone call from Professor M.'s secretary reminding me of committee meeting at four o'clock today.

3:35 Draft letters responding to various nonstandard inquiries about graduate program.

3:50 Pick up key to conference room reserved for committee meeting. Unlock room.

4:00 Return to room for meeting of steering committee for Program in Science, Technology, and Human Values. Only one faculty member and two students there. Other members arrive five to ten minutes late. Director of program brings sherry and paper cups in grocery bag and meeting begins. Last Friday's program with Freeman Dyson reported to have been great success, though low student attendance at luncheon noted. Next week's program and ways of attracting more students discussed. Request for funds to help support a summer course presented. Since program is on austerity budget, request tabled. Meetings with administration scheduled for later in semester may improve budget prospects. Discuss million-dollar proposal just submitted to major foundation.

5:20 Meeting adjourns. Lock room.

5:30 Go to meeting of student chapter of American Society of Civil Engineers. President of society asks if I would be willing to give talk to group later in semester. Agree, though topic is undecided. Think about subtitling talk on Crystal Palace, "Crown Jewel of Victorian Engineering."

5:40 Program begins. Film of construction of world's largest earth dam being shown, compliments of father of civil engineering student whose given name is Bridge. Bridge introduces film and identifies name of father's company on large earth-moving equipment. Watch film and mentally compare magnitude of construction management task with that of Crystal Palace and New York Convention Center.

6:10 Film ends and there is short discussion before pizza arrives. Everyone eats.

6:35 Business meeting begins. Society president announces student-faculty picnic scheduled for Sunday, October 9, and other events planned for semester. Asks if anyone is interested in working on concrete canoe project. Someone suggests putting sign-up sheet on mailroom door. Field trips are discussed. Secretary and treasurer say few words about membership in national society and dues.

7:10 Meeting ends.

7:15 Pick up briefcase from office and begin to walk home. Meet some graduate students and walk along with them until they turn off to play pool in University Center. Walk on alone and think about book proposal.

7:40 Get home and take off jacket and tie. Compare pizza I had with what Catherine and children had. (Theirs was better by three-to-one vote.)

7:55 Talk to Karen about her physics course and whether they have built and tested their toothpick bridges yet.

8:25 Play a video game of football with Stephen. Get beaten 101–0. Play another game and do a little better.

9:15 Read Monday's *Times*.

10:20 Catherine is going to read in bed to prepare her class for tomorrow. She wants to know if she should leave word processor on. Say thanks, but I don't expect to use it tonight.

10:30 Go to desk and look through correspondence on book manuscript completed last summer. If Princeton is not interested in it, who is? Make list of potential publishers to look up in *Literary Market Place.*

11:10 Go over ideas for book on structural failures. Think about themes that would unify everything into single, coherent whole. Dismiss previous attempts that seem now to be too abstract and theoretical. Continue to think in terms of book accessible to wider audience than just experts in fracture mechanics and structural engineering, but with enough meat to make them want to read it also.

12:40 A.M. Little to show for hour and a half of doodling over vague idea. One page of disconnected notes. Pack briefcase and go to bed.

Thursday, September 29

7:10 A.M. Get out of bed after several resettings of alarm. Hear everyone else already up and about even though it is raining.

7:15 Go over in shower what is to be covered in today's class.

7:25 Think about mechanics of shaving. Razor blade must be designed to cut whiskers but not skin. Think about quality control and tolerances associated with manufacturing what must be billions of blades per year.

7:30 Look for shirt-and-tie combination students have not seen. Find some but decide they are unattractive. Select familiar combination that feels comfortable.

7:40 Eat quick breakfast over paper and notice today will be wetter and warmer than yesterday.

7:55 Catherine returns from driving Stephen to school. Have second cup of coffee and talk about possible plans for next summer. University's schedule of courses for summer school now being prepared. Last summer was very productive (a book manuscript and several technical papers completed and submitted for publication), and I would prefer to write again next summer rather than teach. Catherine has been invited to teach at a writers' conference. Karen would like to go to Paris, Milan, and Berlin. Stephen would like to return to Camp Celo in the Great Smoky Mountains. We say we will have to plan our summer one of these weekends.

8:25 Since it is raining, Catherine drives me to campus.

8:35 Agree to be ready to go home about 5:15.

8:40 Check mailbox on way to office. Find memo to graduate students. Proofread and initial and leave with note in chairman's box asking if he sees any new problems being raised by the memo.

8:50 Prepare for continuum mechanics class. Decide to present different argument than textbook's to prove symmetry of stress tensor. Go over development of argument. Select problems to assign for homework. Begin to look ahead to next chapter on Mohr's circle and principal stresses. Plan to cover that before test.

9:50 Interrupted by phone call from dean's office. Associate dean wants to know if I think we should organize evening program to explain graduate school opportunities to our own seniors. Tell him seniors seem already to know all about opportunities and last year's program was very poorly attended. Tell him, besides, I am no longer chairman of Graduate Studies Committee, and refer him to this year's chairman.

10:00 Department chairman brings by memo with suggestion to make one sentence simpler in structure. Otherwise, he thinks it should put the issue to rest.

10:05 Give Mrs. H. final version of memo. Pick up campus mail.

10:15 Graduate student from elasticity class stops by with quick questions about homework problems.

10:35 Go to continuum mechanics class. Review concept of equilibrium equations and present derivation of symmetry of stress tensor. Emphasize point that we have six components of stress governed by only three differential equations. Introduce concept of boundary conditions and give examples. Explain that solving for stresses in one coordinate system gives information about stresses in all coordinate systems because stress is a tensor. Preview concept of principal stresses and look ahead to obtaining other equations. Explain that we must expect to have some kind of restriction on stresses to distinguish one kind of material from another. Argue that if we had block of steel and identically shaped block of rubber, both painted identical black, and if we exerted same boundary forces on each, equilibrium equations could not distinguish between their responses. Thus we have to express somehow what is in the black blocks. Run out of time. Make assignment. Dismiss class.

11:55 Wash chalk off hands. Try to brush it off tie. Think about writing a whimsical piece about the hazard of white lung disease for the teaching profession.

12:00 M. Leave note for teaching assistant telling him what problems were assigned this morning. Find message that someone from Lehigh University called about needing illustrations "ASAP" for paper to be used in the first issue of new journal on fracture mechanics. Glad to hear in this indirect way that paper I sent in response to invitation extended early last summer has been accepted. Will call after lunch hour.

12:05 P.M. Return to office with *Chronicle* to eat quick lunch. Find cartoons best part of day's paper. Look for original drawings for article for fracture mechanics journal. Find only ten of twelve. Locate good Xerox copies of remaining illustrations.

12:20 Find mail not yet distributed. Ask colleague in hall his experiences with programmable hand calculators. Want to purchase one that could be used in lieu of microcomputer to solve such problems as transcendental equations. Leave note for Mrs. H. to get prices on several models.

12:40 Prepare for elasticity class. Go over notes on problem of circular hole in plate. Plan to spend two or three days on it and related problems.

1:15 No first-class mail today. Find note from colleague attached to program from meeting he attended. Titles of papers I asked him about are noted to be collected in a symposium volume.

1:25 Go to engineering library and suggest to librarian that he order symposium volume.

1:35 Call Lehigh and tell secretary of journal editor I will send glossy prints of all twelve illustrations by Monday. Ask for letter of acceptance of paper by journal.

1:45 Leave Mrs. H. illustrations to be photographed. Tell her it is rush job. Go upstairs two at a time.

1:50 Arrive five minutes late for elasticity class. Introduce problem of plate with hole and discuss the complications caused by mixture of circular and rectangular boundaries. Concentrate on geometry around hole and formulate boundary-value problem in polar coordinates. Use principle of superposition to decompose problem further and solve employing stress functions. Realize near end of lecture that book's solution is not completely general and tell class we will begin next session with a discussion of how solution could be made more general.

3:00 Go to meeting of Engineering Faculty Council. Agenda includes vote on recommendations for tightening requirements for a student to remain in school. It is agreed that requirements should be tightened for freshmen since experience has shown that those who do poorly early on tend to get into more and more academic trouble later. It is not considered a favor to prolong the inevitable. Other items on agenda include approval of new courses. Two course proposals are inadequately documented and must be resubmitted. Another meeting is scheduled for next week to continue unfinished business.

4:10 Return to office and make note to prepare paperwork to get new civil engineering course into next graduate bulletin.

4:15 Look at elasticity text to see why authors did not solve problem more generally. See the difficulty more clearly now and decide to elaborate on it in next class. Look ahead to problem of plate with crack and decide to look up original paper.

4:30 Go to engineering library to get 1957 volume of *Journal of Applied Mechanics.*

4:35 Put feet on desk and read Irwin's paper. Find that paper following it deals with strip with off-center circular hole. Serendipity! Decide to copy parts of that paper to illustrate complications involved with a natural extension of what we are doing in class.

5:10 Karen comes to say Catherine is waiting in car. Karen writes on blackboard responsibilities she has for next drama production at school: co-director of set decoration, assistant stage manager, house manager. Congratulate her. Go to car.

5:20 Have a beer with Catherine. Look through day's mail: Tuesday's *Times,* several letters from publishers replying positively that they want to see Catherine's new manuscript, several letters from writers, several bills.

5:50 Stephen comes home from playing basketball. He tells about playing soccer at intramurals after school and football in gym. He tells us he did well on his math test also.

6:10 Dinner is chili Catherine made for yesterday—before I told her I would be eating pizza with students after movie.

6:30 Watch evening news. Learn Congress has passed legislation allowing Marines to stay in Lebanon eighteen more months.

7:50 Wake up after dozing before television.

8:00 Play several video football games with Stephen. Talk about what he does in computer club after school.

8:40 Ask Karen how her toothpick bridge did in physics lab. Find it supported eight kilograms before collapsing. Her teacher's held up beyond twenty-two kilograms. Ask Karen if she will do some drawings to accompany "Mr. Chips" article. Says she will do them this weekend.
9:00 Say good night to everyone. Read Tuesday's *Times.* Enjoy article on innovative airplanes in "Science Times" section and piece on scene in New York Yacht Club while America's Cup was being lost in Newport.
9:30 Watch "Cheers" on television.
10:00 Read several chapters in Giedion's *Space, Time and Architecture.* Note page on which he asks question that serves as point of departure: "Are the methods which underlie the artist's work related to those of the modern structural engineer? Is there in fact a direct affinity between the principles now current in painting and construction?"
11:35 Go to desk and get back to developing outline for book on interrelation of structural success and failure. Begin to converge on unifying theme. List in separate columns examples of failures, successes, and both.
12:20 A.M. Turn off word processor Catherine had left on for me. It is too late to begin writing prospectus for book that is still not completely firm in my mind.
12:25 Return to desk and make more notes for prospectus. Focus on familiar, mundane examples of failures: shoelaces, light bulbs, etc. Resolve that book should be accessible to the layman yet interesting to the structural engineer.
12:55 Check to see if children's alarms are set. Go to bed.

Friday, September 30

7:10 A.M. Hear Catherine going out to get morning paper. Get up after having slept ten minutes beyond alarm's first buzz.
7:15 Think in shower about book prospectus. Had hoped to get

it finished and out to publishers first week in October. Think about what publishers to query.

7:25 Leave empty razor-blade package out as reminder to get more.

7:30 Decide against wearing madras tie and seersucker jacket since weather is getting cooler and it is almost October.

7:35 Have breakfast and read morning paper.

8:00 Leave for campus noticing it is warmer than forecast.

8:25 Stop at main library to look up several books in master card catalog. Get call number for Henry Adams's *Mont-Saint-Michel and Chartres,* thinking it will have something to use in book on structural failure and success. Have no luck finding book on the Crystal Palace whose title I cannot remember.

8:30 Before going into stacks, check new-book shelves and find book on Wallace Stevens. Open to introduction and find interesting quote of his telling his wife to keep his assembling a collection of verses a secret. Take book to table to fill out charge slip and—serendipity!—find copy of *Forbes* with cover featuring article on technology of baseball.

8:40 Descend to subbasement to make copy of *Forbes* article.

8:45 Take elevator up to fifth level to get Adams book. Note other interesting books among the 914.4s and vow to return to browse.

8:55 Go to carrel to look over Stevens book more closely. Decide to check it out and fill out charge slips for it and Adams. Think about how many hours were spent in carrel last summer, and look ahead to spending some more when classes are out.

9:10 Go to current periodicals section. Walk through English aisle and look through September issue of *Poetry.* Elsewhere locate *Science, New Scientist,* and *Scientific American* and take to comfortable chair.

9:45 Check out Adams and book on Stevens.

9:50 Look over drafts of manuscripts displayed in library lobby.

Note that practice makes, if not perfect, certainly better.

10:00 Enter School of Engineering and check mailbox. Find memo to graduate students has been distributed. Hope it does not precipitate trouble the way roster precipitated it. Take campus mail to office.

10:05 Open campus mail. Find note from director of Duke Press acknowledging my note re Freeman Dyson and possible manuscript from a friend of his in Princeton. Also find copy of contents page from latest issue of *The Engineer.* Note article on North Sea offshore structures and put on pile where I know I will find it in November, when *Technology Review* will be editing article on offshore oil platforms and will be looking for updates and sidebars for manuscript prepared last spring.

10:10 Get call from university bookstore setting up meeting of visiting faculty committee.

10:15 Call MIT Press re manuscript I mailed in August. Editor is not in his office. Leave message asking if manuscript arrived and if it is being reviewed.

10:20 Draft letter to be ready to accompany prints of illustrations to Lehigh University as soon as they are ready.

10:25 Check time of paper in Houston. Decide to put off preparing transparencies for talk until week before meeting.

10:30 Look over computer program graduate student has left in mailbox. Write note saying he should go ahead and run cases we had discussed.

10:50 Professor A. comes in and says today is last day to list in official schedule courses to be taught next summer. Tell him I have no plans to teach since I will have research contract and want to write.

10:55 Telephone call from student re graduate school. Explain procedures for applying for admission and financial aid.

11:05 John Fast stops by between classes. Tell him about finding

Forbes article and noticing two cases of "technological illiteracy" in cover illustration of baseball story. Discuss article and various and sundry other topics. Suggest he draft letter to the editor pointing out errors confusing momentum and energy and drawing of piece of chalk broken in a virtually impossible pattern on *Forbes* cover.

11:30 Draft letter to Freeman Dyson re his visit to Duke and our discussion re Duke Press. Draft letters to students who have asked specific questions about graduate program.

12:15 P.M. Deliver various items to mail room. Pick up mail and *Chronicle*.

12:20 Read letter from editor at John Wiley re galley proof of my chapter in book on offshore structures. Proof should be expected in about two weeks and must be returned promptly. Read through elaborate instructions on how to mark proof and how to prepare index. Surprised at detail and at work expected of author. Did not realize all this would be involved when I agreed to write chapter almost a year ago.

12:40 Read *Chronicle* while eating usual lunch. Think about the absence of luncheon meetings this week and almost daily ones next week. Realize paper is very large edition for Parents Weekend. Front-page story on North Carolina drinking age being raised to nineteen tomorrow. Read articulate letters to editor from engineering students and try to remember how much higher than the university average are the verbal SAT scores of Duke engineering students.

1:05 Go to engineering library. Check new-book shelves. Find book published in New Delhi on large plastic deformations. Discover nothing new in chapter most relevant to my research but note that sources of familiar illustrations are not acknowledged. Find another book, *Global Stakes: The Future of High Technology in America,* with a chapter that contrasts education in the humanities and in engineering. Fill out charge slip.

1:35 Check mailbox. Find plenty of letters to be signed. Message that registrar's office wants me to call re a graduate student's registration.

1:40 Call registrar's office. They have no record of student's dropping course he was registered for and adding courses he claims to be attending.

1:45 Try to call student in question. Get no answer in his office. Write him note to see me re his registration.

1:50 Look into elasticity problem of plate with hole whose radius is not necessarily small. Begin to work through algebra for general case and soon realize why restricted case is always considered in textbooks. Decide against carrying calculations out to end. Put aside until next week.

2:30 Sketch out paper on response of beam with crack and load at arbitrary locations. Anticipate results of computer program and decide to start writing introduction and developing analysis before graduate student brings computer output.

2:45 Check to see if chairman is going to seminar to be given by dean at 3:00 P.M.

2:55 Stop by chairman's office on way to seminar in Physics Building.

3:00 Arrive just in time for dean's seminar on stability and chaotic behavior in aeroelastic systems, being presented to applied mathematics group. Interesting seminar with plenty of questions from audience.

4:05 Walk back to engineering school. Hear cheers for egg-drop contest taking place on lawn.

4:10 Find large crowd gathered around site of contest. Find good vantage point on hill. Helicopter-like device is being dropped from roof of building. Egg breaks on impact with concrete patio. Crowd moans.

4:20 Walk over to keg and try to get some beer. A former student of mine has control of spigot and fills my cup.
4:25 Return to vantage point and watch crowd move back as cinder block with egg attached is dropped from roof. Watermelon encasing egg follows. Some serious entries are very ingenious and land with egg intact. Unbroken eggs are smashed against wall to show they are not hard-boiled. Crowd cheers as glider with egg pilot soars over everyone's head but then moans as entry crashes into side of building.
4:50 Catherine comes around corner. She has manuscript to copy but Xerox machine is broken.
4:55 Last egg is dropped. Keg is empty.
5:05 Catherine and I stop by office to pick up my briefcase.
5:15 Pick up children at University Center, where they have been playing video games.
5:25 Park car in carport and swear we will not go out again tonight.
5:30 Have drinks while dinner cooks.
6:15 Dinner.
6:45 Tune in evening news.
7:15 Turn off television and go out to catch football with Stephen.
7:35 Watch "Family Feud" with Stephen.
8:00 Discuss plans for tomorrow. Stephen has soccer game at ten o'clock and there is an engineering seminar at the same time. There are also Parents Weekend events after the seminar, but I am not sure whether to go. Football game has been rescheduled for late afternoon so that it can be carried on TV, and that messes up the early evening.
8:20 Read Wednesday's *Times,* October *Technology Review,* and chapter eight of *Global Stakes.* Find a description of "the two cultures" that relates the story of C. P. Snow's Rede Lecture by talking about "Newton's Second Law of

Thermodynamics." Make a mental note for technological illiteracy file.

10:55 Everyone else asleep. Go to study and work on prospectus for book on structural failures and successes. Find notes of previous nights read well. Prepare narrative summary to go with chapter-by-chapter outline and decide on working title, *To Engineer Is Human*.

12:50 A.M. Read through summary and outline one more time. Decide it is ready to be typed into word processor. Determine to do that at first opportunity, but not tonight.

1:10 Check Stephen's alarm to be sure it is not set too early for soccer game.

1:15 Set my alarm for 8:30, early enough to make either the seminar or Stephen's game.

1:20 Go to bed figuring time I spent with engineering students and with Stephen this week, and lean toward going to soccer game.

At Home

15·Time Piece

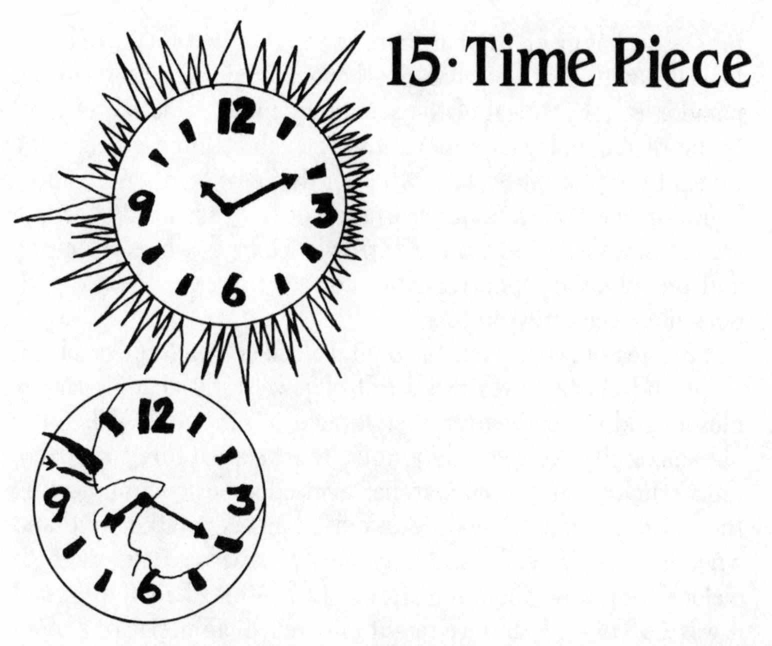

Once upon a time, when clocks and watches of mechanical design could stop and could be stopped by bullets, and when shrewd sleuths found clues to murders and to other crimes in which the hands were frozen at 2:06 or some such incriminating hour and minute, a wag observed that even the most broken of timepieces was correct to the second exactly twice each day, no matter where the hands might be frozen. This trivial truth, that time will pass the stopped clock as surely as the steady tortoise will pass the resting hare, has a curious counterpart in the digital electronic timepieces of modern technology: no matter how accurate and reliable they may appear to be, even those electric alarm clocks that hold batteries in reserve so that we not be failed should the power falter the night before an important meeting are *wrong* at least once and possibly twice each day. Why this is so will become clear in due time.

For all the opportunity that the new technology of microprocessors has given us, that to make watches that finally tell time logically with numbers has not been seized. And this curious paradox is not atypical of the seemingly inescapable tautological lesson of technological innovation: the more things change, the more of the past they retain, whether that past be right or wrong, logical or illogical, rational or irrational. In other words, technological innovation, like other forms of change, while promising and providing an opportunity to correct the errors of the past, does not necessarily do so.

The greater accuracy of the digital clock or watch full of silicon chips and crystals over those full of jewels and gears is largely illusory and more a matter of style than of substance. When the old analog devices were ubiquitous, few people other than saboteurs synchronizing their watches worried about communicating the "exact" time. When we asked strangers for the time and when we gave it, we by and large would say it was "almost three o'clock" or it was "about ten after eight." We seldom, if ever, said it was 2:58 or 8:11, and we might only tell an amnesiac or a blind man that it was before or after noon. Of course, if we wanted them to be, our watches could be so correctly set that the second hand reached its zenith just as the tone on the radio announced the beginning of our favorite show. And we often developed a curious sort of symbiosis with our mechanical parasites, winding them up on a diurnal cycle as dependable as anything else in our lives.

The old timepiece was capable of involving all of our senses with the passage of time. Our *eyes* could watch the continuous flow of time as second, minute, and hour hand moved around the face beneath the crystal as steadily as the heavenly bodies once moved with the music of the spheres. On the larger clocks we could see the minute hand move inexorably around the dial as surely as we can see the moon move across the night sky. And we knew the hour hand was moving as surely as do the stars. Our *ears* could hear the *tick-tock,* the uniform digitization of time, and the more important the passage of time seemed to be, as in the middle of a sleepless night, the more loudly we heard the innards of the clock wearing the sharpness of the gears dumb. Our *fingers* felt the main spring tighten under the action of the winding stem as surely as the force that through the green fuse drives the flower drove Dylan Thomas's green heart. Our *nose* could smell our very own perspiration upon our watch bands and our *tongue* could taste the salt from our pores upon the metal back of the watch that we married to our wrist for life. According to Daniel Boorstin, one 17th-century Frenchman went so far as to replace the numerals in his bedside clock with twelve different spices so that his senses of smell and taste could tell him in the dark where the hour hand pointed.

The newer kinds of watches that we buy and discard with little emotion and no regret have no such universal sense appeal. They are silent, tasteless, and odorless. They are so light that we hardly work up a sweat wearing them, and they of course require no coordination of thumb and finger to wind. If they excite any sense, it is the visual, but not because any of their parts are moving gracefully, as the moon or clouds across the sky. No, if anything, the sudden change of the minute or the frantic counting of the seconds in digital quanta is visually abrupt and unnatural. And the design of the digital watch or clock and its awkward crew-cut numerals seem not yet to have reached an aesthetic equilibrium. If the Museum of Modern Art has enshrined digital timepieces, it is not because of their numbers but because of their (mat black) cases. And just as a beautiful-looking typewriter may make an ugly typographical impression on a piece of paper, so an attractive digital clock can be an abomination when telling us it is 12:21. While the numerals *1* and *2* are juxtaposed for the hour as they should be expected to be, the *2* and the *1* for the minute are spread apart in as ugly a typographical faux pas as letterspacing lower case. This is no fearful symmetry.

It is possible to set a digital watch as accurately as any old analog, of course, but I have not found the wearers of the newer chronographs to be so fastidious as the second-hand setters of old. I recall many a time sitting before my radio, my thumb and pointer holding the watch stem cocked just right so as to keep the second hand poised for the start of a race of seemingly countless laps around the track of time. With the tone I would release the stem and push it in on a timepiece whose minute hand I had already set exactly on the mark of zero seconds. While such synchronization should be possible with (and seems to be demanded by the technology of) even the most inexpensive of digital watches, it does not appear to be done. My son plays a game with his digital stop watch. He gets its seconds going and tries to stop it on an even minute. He freely admits that it is a game of chance, and he rarely takes the time to set his watch to the second. And neither do adults, it seems, for I have been in many a meeting when digital alarm watches around the table have beeped or played "Dixie" not all at once but over a matter of minutes on either side of the hour. Hence all those minutes and seconds changing in step with the pulsing of quartz crystals are only giving the illusion of accuracy. They are no more correct than the change given by a chimpanzee.

But if digital clocks *can* be set to start off from 12:00:00 when the tone on the radio signals noon, how is it that they are wrong at least once each day? The answer lies not in numbers but in words—and in Latin words at that. Digital clocks not only tell the time so that we do not have to interpolate minutes and seconds between marks on the face, they also show whether the time is A.M. or P.M. Unfortunately, these abbreviations have come to be used so carelessly as labels for morning and afternoon that their true meaning has been frequently forgotten. The *m* in these abbreviations stands, of course, for *meridies,* which is Latin for mid-day or noon, and A.M. stands for *ante meridiem,* or "before noon," and P.M. for *post meridiem,* or "after noon." (The change from *-dies* to *-diem* is required by the declension of the Latin noun.) Hence noon is properly written 12:00 M. Yet every digital clock that I have seen shows 12:00 P.M. (or 12:00:00 P.M. if it tells seconds) for noon—a flagrant error in Latin and in chronology.

It is easy to imagine how this error might have arisen. First, it is a common confusion as to whether to write noon as 12:00 A.M. or 12:00 P.M., and many an itinerary, program, or schedule will employ a curious mix of Latin and English by listing morning hours as "A.M.," noon as simply "Noon," and afternoon hours as "P.M." A second source of the error (or neglect) in programming digital clocks may have a more or less excusable reason, depending upon one's point of view. The technological problem of having a clock show 11:59:59 A.M. for the second before noon, 12:00:00 M. for noon, and 12:00:01 P.M. for the second after noon would require a much more sophisticated microelectronic chip and display than is inside the plastic watches. Such chips and displays could be made, but the additional cost would remove the attraction of the cheap and "accurate" digital watches that are nowadays given away as premiums. Whether or not a conscious decision was ever made by digital watch designers to sacrifice correctness for economy may never be known.

The question of how correctly to designate midnight is more problematical, and it can be argued that the witching hour is both 12:00 A.M. *and* 12:00 P.M. This ambiguity arises from the traditional arrangement of the duodecimal numerals in the circular pattern without obvious beginning or end. While this continuity may appear to be attractively analogous to the diurnal flow of time, which itself has no perceptible beginning or end, it is as difficult to argue that the sequence of the hours "begins" with 1:00 as it is to argue that it "ends" with 12:00. Indeed, since 12:30 always precedes 1:30 in time-telling, it is logical to say that the sequence of hours is 12, 1, 2, 3, . . . , 9, 10, 11. Whether midnight is the beginning of the sequence, i.e., 12:00 A.M., or the end (i.e., 11:60 P.M. = 12:00 P.M.), is not at all obvious, to me at least.

The twelve-hour clock as we have used it thus must share some of the blame for abusing Latin, and the inconsistencies of the long-standing system become obvious when we try to digitize time in a consistent manner. If everything is really reckoned from noon, i.e., 12:00 M., then it certainly makes sense to have midnight, which is exactly twelve hours *after* noon, as 12:00 P.M. But after midnight we cease to count *after* noon and begin to look forward to the next noon, so that the minute after midnight is not generally thought of as twelve hours and one minute after noon, i.e., 12:01 P.M., for that would be ambiguous, but it is rather 12:01 A.M. Yet it might also be argued that midnight is exactly twelve hours *before* the next noon and hence midnight could properly be labeled 12:00 A.M. The reason midnight presents special problems to a consistent telling of time with numbers is that in fact the *12* on our clock faces really shares the space with and may be thought to cover up a *0*. When the hour hand points between the *12* and the *1* we could really say 0:30 A.M. (read "oh-thirty in the morning" or "half past midnight") or 0:30 P.M. (read "oh-thirty in the afternoon" or "half past noon"). Midnight is such an ambiguous hour precisely because it is *both* 12:00 P.M. and 0:00 A.M., or, with the understanding that *12* is another designation for *0*, both 12:00 P.M. and 12:00 A.M.

Now it is not really any harder to say "oh-thirty" than "twelve-thirty," but the habit of telling time by calling out the last hour numeral that the little hand has passed has driven "twelve" into our chronic response. We do not see a zero on the clock face, and we do not hear it in anyone's civilian vocabulary. Yet the use of *12* for the hour between noon and 1:00 P.M. (or between midnight and 1:00 A.M.) has made our otherwise symmetrical system of time-telling curiously skewed about the one-o'clock position. While the more numeral-oriented "faces" of digital timepieces may have made it natural to use *0* in place of *12*, usage and custom seem to have prevailed over any sense of numeracy.

It is really the words for the numbers that have come to be adopted for time-telling, and the unimportance of the numbers themselves may be seen in the design of many an analog clock or watch face without any numbers at all. In the extreme case, a precious stone might mark the position of noon (and midnight), and there might not even be divisions for the other hours. The digital timepiece is no ticks and all numbers, however, and hence it makes us think about time-telling as numbers pure and simple, and therefore it suggests that the telling of time should employ numbers as numbers are customarily used—and that means as measures as well as signifiers.

There is another curious skewing of our method of time-telling, for we reckon before and after noon in inconsistent ways. Times after noon, understanding that times like 12:30 really mean 0:30, occur in ascending hours, properly indicating the temporal distance away from (or after) noon. Thus 6:00 P.M. is three times as far from noon as is 2:00 P.M. The measure of time before noon, however, involves an inexplicable inverse use of numbers and (Latin) words. As we progress with the sun through the growing morning hours—8:00 A.M., 9:00 A.M., 10:00 A.M., etc.—we are not moving further away from noon as the increasing numbers in conjunction with the Latin *ante* suggests, but rather we are *approaching* the *meridiem.* How accustomed we get to the inane inculcations of our youth! Oh how proud we are when we learn to "tell" time! Which of us wants to think forty years later how arbitrary, if not downright wrong, it all sounds? (Military time, by reckoning from zero hours on a twenty-four-hour clock, is a rational but not a classical solution to all these problems.)

Time was when the numerically irrational symmetry of the clock face about noon and midnight played an important role in the selling of clocks and watches. In catalogs and in display cases the hands were uniformly arranged to read about 10:10 or 8:20 (A.M. or P.M. never was an issue), and an entire page or window or showcase was synchronized as if to demonstrate that the accuracy of the individuals was somehow greater because of the universal agreement of the ensemble. This habit of merchandising was even the topic of an extended discussion on the editorial pages of *The New York Times* some years ago, and there seemed to be more letters to the editor offering explanations as to why those ubiquitous times were chosen than there are Apocrypha. I do not recall all the arguments given at the time, but if they did not they certainly might have included the pros and cons of a smiling clock face frozen at 10:10 versus a frowning one set at 8:20.

To one with a mathematical as well as a psysiological and a psychological bent, the times 10:10 and 8:20 suggest not only facial expressions but also geometry. These singular times are simply the only two balanced arrangements of the hour and minute hands in which they assume symmetrical positions (about the arbitrary but privileged 6–12 axis) that are neither too rigid- nor too acute-looking. In fact, if my algebra is correct, the hands should be set at exactly 10:09:14 and 8:18:27 to have the angle between them bisected by the (unsynchronized) second hand pointing to high noon (or dark midnight).

To check my memory that these are indeed the way clocks and watches have been displayed, I looked in the catalogs of two large discount chains. And I found a curious arrangement of timepieces. In one catalog all analog clocks were consistently set at about 8:18, but with the second hand rather as obstreperous as a cowlick, while all analog wristwatches were set at about 10:09, again with an errant second hand. Perhaps even more interesting in this catalog, digital clocks and watches showed a variety of times, including 12:00, 12:08, 12:13, 1:03, 3:11, 4:03, 4:04, 5:00 (never mind the A.M. and P.M.), and even some showing (numerically very logical but emotionally as dead as Latin) military times like 16:26 and 20:09. It is as if the accuracy of the digital clock or watch is so taken for granted because it displays time as a *number* that it can arrogantly display *any* time (and any number) in an advertisement.

Another chain's catalog flouts traditional display times, and all analog *clocks* in it show the time to be an asymmetrical 7:20, with all digital clocks also showing the same hour and minute, but with disagreement as to whether it is before or after noon. However, when I turned to the pages of *wristwatches,* I found chaos. The analog watches show 10:10, 8:20, *and* 7:20, while the digital watches show almost as many different times as there are models displayed. I do not quite know what to make of this, other than to observe that the digital watch, like the digital computer, seems to be sold on the basis of its ability to display any and all, rather than necessarily the correct, numbers. It is as if it is taken for granted that the numbers one can get out of the electronic labyrinth of open and shut gates are correct—even though one can really get any numbers out that are set or programmed in. And the numbers one gets out of an electronic brain, whether the size of a watch strapped to the wrist or the size of a supercomputer that one is saddled to, can be both right and wrong. And even when the numbers themselves may be correct, the words that accompany or interpret them can be as diametrically opposed as night and day.

What the digital clock and watch teach us is that more information (including more explicit information) is not necessarily more precise or correct information. The *ability* of a digital watch not only to display the hour, minute, and second of time in numerals that change more quickly than we can call them out but also to go beyond the mere numbers to display the A.M. and P.M., is not necessarily an improvement in time-telling. Some have observed, in fact, that the digital display is overly detailed in its information content, and that the analog display of "about" times is much more suited to our sense of time as temporal distance when traveling between one place and another or in timing the termination of one appointment in order to make the beginning of the next "on time." And the digital watch or clock, unlike the old timer, whether with its hands up or down, whether smiling or frowning, is worthless when it comes to providing clues as to when a crime might have been committed. The light-emitting diode and liquid crystal displays of digital clocks and watches no more freeze at an incriminating instant than do their plastic cases stop bullets. The only clue the modern marvels might give to a Sherlock Holmes is that their designers were either ignorant or simply did not or could not take the time to do it correctly. Advancing technology has no right to promulgate error, whether it be numerical or alphabetical, but it can and does all the time.

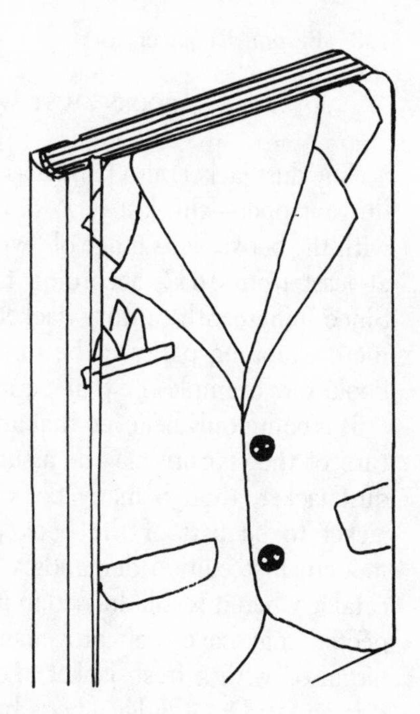

16 · Dust Jacket Dilemmas

Books are the liberated spirits of men, and should be bestowed in a heaven of light and grace and harmonious color and sumptuous comfort. . . .

—Mark Twain

It appears to be a matter of fashion, if not one of etiquette, that today's young trade book does not make its debut in bookstores without wearing a dust jacket. Though it may be gaudy (a loud, bleeding madras straight off the rack and advertised in the Sunday rotogravure) or conservative (a tasteful blazer, tailor-made and available only at Books Brothers), a jacket must be worn. It is as if bookstores, paraphrasing their restaurant neighbors, display this sign in their windows:

BOOKS MUST WEAR JACKETS.

The dust jacket, also known as the book jacket, dust cover, and dust wrapper—this last term to be eschewed, lest it be confused with the bookseller's usage of "wrappers" for paper covers—dates at least from 1832, according to John Carter, the bibliophile. Since nineteenth-century jackets were considered ephemera, mere wrapping paper to be discarded by the purchaser of the book, few examples of that period survive, however.

It is commonly believed that any trade book published after the turn of the century may be assumed to have been issued with a dust jacket, though many book collectors do not consider the jacket to be part of the book proper. Nevertheless, even the modern first edition demands a higher price in the bookseller's catalog when it is still dressed in its original jacket. But books, like people, may have their provenance, and their worn shirtsleeves, disguised with a fresh jacket. To all, therefore, Carter provides this caveat: Don't judge a book by its dust cover. Indeed, my own copy of his *ABC for Book-Collectors,* a "Fourth edition, revised," according to the copyright page, provides an example for the lesson: its jacket advertizes it as a "Third Edition, Revised."

When a book's jacket is original and correct, however, it often reveals a lot about where the book has come from and what its heritage is. The prepublication praise by the author's contemporaries, his youthful portrait and biography, and the book's original price are among the unique pieces of information that an original jacket may not cover up. Often these things are surprising in retrospect, especially the price. The jackets of books in my own library prove to me that I am not dreaming when I remember that a book of poems once sold for $2, a book of essays for $2.50, a first novel for $3.95, a major biography for $5.95, and an engineering textbook for little more than $10.

But just as those are not today's prices, so the jackets on which they appear are not today's styles. Even if it is after the latest fashion of pop, op, or co-opt art, the contemporary book jacket is basically an advertising medium, a slick wrap-around billboard. It is often a masquerade, a facade, a cosmetic. But as cosmetics can make both the beautiful grotesque and the plain attractive, so the book jacket can be out of harmony with the universe it envelops.

While it is true that many books are dressed in jackets whose designs are hand-me-downs from other children of the author, or cousins, especially successful older ones in the publishing house, there is many an only child in the book population and new families are constantly being started. And it is often these for which the dust jacket is tailor-made.

The book designer and jacket designer are seldom the same individual, however, and when one is an artist often the other is a salesman. Nevertheless, there are perfect fits, when the book and its jacket are as one good design, and therein lies the problem for the book collector. Does he or doesn't he discard that jacket before stooping, reaching, or climbing his library steps to shelve another example of the book-publishing art?

If shelf space is at a premium, as it is in many a home of bibliophiles and -maniacs, a cold, pragmatic approach might lead one to consider a cost-benefit analysis of the dust jacket question. I recently borrowed a machinist's micrometer and measured the thickness of the dust jackets on a sample of books—including works of fiction, history of science and technology, design, and construction engineering—shelved close by my desk, and I found an average thickness of six-thousandths of an inch. Each jacket with its flaps adds four times this thickness to the thickness of a bare book, which for my sample averaged about an inch. Thus to shelve each book with its jacket requires an additional 2½ percent of shelf space over what would be required to shelve unjacketed books. Or, looked at from another angle, by discarding dust jackets one can shelve about 2½ percent more books in the same amount of space. This is equivalent to an extra book for every 40, and 25,000 books—a fair-sized branch library—for every million volumes in a large university library.

But such numbers might be meaningless for the private or amateur collector of books. And as he may not be able to decide on the basis of purely quantitative evidence whether to shelve his books with or without their jackets, so he often cannot easily decide whether to remove from its jacket—even temporarily—the book he is reading. To remove the jacket exposes the cloth to the secretions of his palms and fingertips, which might be dipping into salted peanuts or potato chips. Yet to leave the jacket that he wishes to preserve on the book during reading can cause it to become greasy, its edges to tear or curl, and its gloss to dull permanently. To leave or not to leave a dust jacket on is as difficult a decision as that which a jacketed man who perspires freely faces at a hot summer dinner in a house without air conditioning, and it is among the higher dilemmas of reading.

Ironically, in the library of fine bindings, where if readers nibble it is not from bowls of greasy snacks, books wear no jackets at all and go in shirtsleeves. But what shirts—soft calf tooled and embroidered with the brightest gold. And just as the saddle must not overwhelm the show horse's color and conformation, so a flashy jacket could never be allowed to compete with morocco. So the books are shown bareback. In the library of fine bindings, therefore, the dilemma is as real as a unicorn.

In the public or institutional library the dilemma presents itself as inexorably as a one-horned rhinoceros. Whether prompted by space considerations or not, the librarian seems to follow an invariant policy to either discard all jackets or to attach irrevocably each jacket to its book with plastic and tape. I know of only one major research library that not only leaves its jackets on but also leaves them unattached, like a European with his jacket draped across his shoulders but his arms not in its sleeves. This library even goes so far as to put the call number on the jacket *and* the book's spine.

But who has not been frustrated by a library's singlemindedness? Who has not wondered if the missing jacket contained a photograph of the poet one has just discovered in the university library? Who has not folded back the cover boards of a tightly jacketed book from the public library to look into a dark tent for the manner in which the book's title has been placed on the spine or for an author's signature or monogram imprinted into the cover cloth? These are among the pleasures of holding books in one's hands.

When I inspect the cloth beneath the taped-on dust jacket of a library book, the jacket and the boards form a diamond, and the pages of the book fall down, hang apart in signatures, and fan out. The forgotten bookmarks of previous borrowers, newspaper clippings, reviews, grocery lists, lists of unfamiliar words, overdue notices, and perhaps some paper money—all fall out, and the pages of words repel each other like the leaves inside a Leyden jar. I look into the diamond mine, flexing the acute and obtuse angles, jockeying my vision through the tunnel to catch the light on the spine. I inspect the spine and the cover and wish I could see the veins of gold more closely, could feel with my eyes the texture of the cloth. I wish the jacket were not taped to the endpaper, but at the same time I am glad that it is, because the cloth is fresh and clean and the gold- or silver-stamped title is shining in the dark, unfaded by the sun and uneroded by the sweaty palms of —how many readers? Let us thank those librarians who do keep the dust jackets taped to their books. But let us also thank those who give us the straight cloth, even though they do discard the jackets after their brief display on the new-book bulletin board.

Yet the dilemma remains in one's personal library. Is the dust jacket to be a part of the book? If the design of a dust jacket is primarily an advertising rather than an editorial decision it would appear that the jacket *is* a mere wrapper, packaging to be discarded. The function of the dust jacket would be to sell the book on the bookstore shelf. Period.

In graduate school, where book-collecting reaches fanatical proportions, I knew someone who wanted to take off his dust jackets and keep them too. He laid them out flat, alphabetized, in map drawers, and he designed bookshelves that would sag as little as possible under the weight of his sunbathing books. He used a textbook on the strength of materials to calculate the deflection of book-laden boards of pine supported at two points on bricks, and he used calculus to locate the bricks for the minimum bow of the board. When the optimized shelves-cum-room divider were erected, he waxed them so that the books might be slid from their slots with a minimum of friction.

Other friends worried more about the horizontal than the vertical placement of their books. Some numbered and cataloged their books, and one couple even had a box of index cards so that a visitor might browse through the library dispersed throughout the various rooms of the apartment without leaving the easy chair in the living room. Another couple followed a strict alphabetization by author and circled an entire room with floor to ceiling bookcases that juxtaposed the *Z*s with the *A*s. Other less serious and certainly less reverent friends arranged their books by kind—paperbacks horizontal and hardbacks vertical—by size or by color. In this latter arrangement the ROYGBIV spectrum of physics circled the room like a rainbow until violet clashed with red. Since graduate school days I have learned that one can hire, at least in New York, free-lance librarians to use these and even more exotic schemes to organize book collections.

But even as my friends reached no consensus on the status of the book or its jacket, so the dilemma about its disposition remains. Unfortunately, or fortunately, perhaps, the jacket often supplements the book with information of great interest to the reader and bibliographer. Although many books now include something "About the Author" in their last pages, this is by no means always the case, it being subject to an editor's or a book designer's preferences, space limitations, multiples of sixteen, and other ad hoc influences. A dust jacket photo of an author, almost never repeated inside the book proper, presents a further dilemma to one who wants to keep the photo but who also wants to shelve his books barebacked. And the reader who wants to preserve some unique information about an author, without cutting up the jacket and pasting clippings on the end papers of his book, is obliged to leave the dust jacket on or (if he would not crease the round folds of his jackets) maintain a separate and possibly oversized file of separate covers. Most find both alternatives undesirable at best.

I have discovered no satisfactory solution to the great dust jacket dilemma. I do keep my books in their jackets, which do not strike me as ephemera. In fact, a jacket is often made of finer quality paper stock than the pages of the book itself, and often a well-fitting jacket disguises a disfigured novel, its perfect-bound backbone permanently misshapen after a single reading. However, I do, very frequently, remove from its dust jacket an arm of the book I am reading to see and feel what the book is *really* like. I want to see the cloth, the spine, and a full spread of end paper, and I want to touch the naked parts. Such experiences can be as pleasurable as reading favorite passages over and over again.

Book designers, I like to think, have agonized over the texture and color of the cover cloth as much as they have over that of the end papers and of the pages of typography on which the literal art is to be impressed. And I want to understand and take pleasure in their so-right choices as much as I do in the mots justes inside. Every manuscript has a color, a texture, a typeface, a page that is more right than any other for it. The designer must bring these out of the manuscript as the sculptor must bring the figure out of the stone. To design a book and to bind it in its proper binding is almost to write that book again.

And I should like to argue that if there must be dust jackets, there must be well-designed dust jackets around well-designed books. For the jackets too I want to fondle with my eyes. I want to savor the graphic art, the blurb précis, enticing and precise, the jacket *portrait,* the author's note, as much for what it reveals as for what it leaves to the imagination. I enjoy having the way of the word strewn with petals of praise by authors I admire. I want to open the book by turning a cover that does not tell me the story inside, but that does hint of the colors of the words to come. I want my book, and I want to read it too.

Though the dust jacket does seem to me to be primarily an advertising medium, I do enjoy well-designed packaging. The beautifully painted eggshell does not diminish the egg. Its shape remains and the egg tastes none the poorer for having been illuminated for children in the Easter grass. The shape of the vase is enhanced by its glaze.

The packaging of many textbooks and children's books has for some time precluded the decision about dust jackets, for the cloth binding and jacket of these books are often one and the same—the book's boards are covered with a washable, printed material that also serves as a jacket. And the paperback industry, in which the book jacket and cover are one and the same, has gone so far as to issue projected million-sellers in a spectrum of colors to cover the spectrum of moods in which travelers fly and in which drug and grocery shoppers approach the book racks that flank the checkout lanes.

Since the book-publishing industry has not yet adopted universally this leisure look in dust jackets, the hardcover trade book still wears the traditional garb. Whether this will change is of no matter; the book owner's dilemma remains for the present. And just as the executive hardly can recognize or be recognized by his associates on Saturday, when they all shop for fertilizer in their shirtsleeves, so many a reader cannot recognize his favorite book without its dust jacket.

17 · A Little Learning

When I riffled through my son's first-grader reader, I was furious at what I saw on page 56. Not only the *O*s, but also the *B*, the *D*, and every letter with a loop that could be, was filled in as black as the ink of the chapter title: WE READ BOOKS. I could not condone this vandalism of a school book, nor could I ignore the boredom of which I thought it symptomatic.

Fortunately, before I called my son to account for this reprehensible act, I noticed that the letters were filled in a little too neatly for most six-year-olds. On closer inspection I discovered that this apparent vandalism was part of the book's design, and the graphic act was repeated on other pages.

This is madness, I thought. Why should a child's first reader even suggest an activity that to many symbolizes dissatisfaction with reading? Were the designers trying to prevent the child's own filling-in of letters? But how could such an anti-intellectual act be given legitimacy in any textbook? Who was responsible for such a bad example in a primer?

My son's reader is published by an established textbook firm, now a subsidiary company of one of the world's largest "word-processing" corporations. A senior author is listed on the title page, and the copyright page identifies, by name, three contributing authors; two consultants, one for linguistics and one for creativity; and two artists. Did none of these object to what I found on page 56?

I studied the book more closely and found it a tremendous disappointment. No wonder the filled-in letters went unnoticed by the book's producers. They, like all the reader's young readers, must have been bored, bored, bored, and paid little attention to what they were reading or proofreading, I thought.

Since this essay is going to be critical, it will not be an infringement of copyright to quote in full the short first story, "At the Park."

> This is Bill.
> Bill is at the park.
> Here is Jill.
> Jill is at the park.
> Here is Ben.
> Ben is at the park.
> Is Lad at the park?

The most interesting thing about this story is the question at the end, and I suspect that most kids expect an answer on turning the page. Do you think they get one?

Following "At the Park" is a story called "The Ducks," in which Ben feeds the ducks something identified in print only as "this." The ducks get a bit excited at this, and the story ends in a frenzy after Ben asks the ducks to "Stop this!" What is this? Who is Lad?

Well, Lad is a dog who apparently runs out of some earlier or later book in the series to endanger the hungry ducks and chase them away. And so the tale wags the dog.

As I read the reader it thus appeared to me that a team of four authors, aided by a pair of consultants, had found little to change but the names in the beginning readers I had used some thirty years earlier. This confused me.

On the one hand, I was glad to see that these educators had not instituted change for the sake of change. After all, if the old ways have been effective, they should be retained, unless there is some good reason to replace them. It was good to see educators and consultants not just making work for themselves.

On the other hand, there seemed to be something wrong in the tone of this book. There was something disappointing. The book was—boring. It made reading seem as tedious as assembly line work. That was it—the book was assembled from standardized words, rather than being custom-built, as books should be.

Although there are over 1,200 total words in the text of my son's reader, about twice the number that have so far been used in this essay, there are only 58 *different* words. No wonder the reader sounds like a broken record. (According to my computer, this entire essay, which I suspect uses a much more restricted and repetitious vocabulary than my typical writing, contains 2,558 *total* words, with 756 different words. That is a ratio of about 6/20 as opposed to 1/20 in the reader.)

I thought how my son and his classmates must be bored stiff with this book, for he and they can already read all of its fifty-eight words in any combination of sense or nonsense. If they have difficulty, I thought, it is in following the story by words alone, as I thought young readers should, for that is near impossible.

When Ben feeds the ducks "this," the young reader must wonder what Ben is holding over his head in the crude illustration. When Ben seems to offer Lad "this," it is just as vague. When Miss Hill, in another story, tries to guess what the children have in a covered box, she asks, "Is it this?" several times, but "this" is only identified in the illustrations of Miss Hill drawing at the blackboard.

This is wrong. In the hundreds of library books I have read with my children, the text has generally been self-sufficient, and it did not rely on the artwork for its referents.

But back to boredom. My son and his classmates have been brought up with "Sesame Street" and "The Electric Company," and they can read individual words left and right and upside down. By the time these children reach first grade they no longer watch these television shows, because *they* are boring to the veteran kids. Now that the television has been turned off (in anticipation of "real" school?), let's not turn off the children to reading with readers more boring than television.

At breakfast my son reads the list of ingredients on the cereal box. On the way to school he reads the highway signs. After school he reads the television schedule to see what is on opposite "Sesame Street" and "The Electric Company." He doesn't need fifty-eight words used twenty times each that can't tell a story without pictures. His reader may keep him from reading. When are our educators going to learn?

So I had it all figured out, and just in time for curriculum night, the first formal encounter of the year between parents and teachers. I didn't know if or how I would confront my son's teacher with my brilliant, critical analysis of the reader she may have selected, so I paid close attention to her presentation of the first-grade subject matter and materials.

She disarmed me immediately by dwelling on the very series of readers whose example I found so disappointing.

She, too, had been surprised that our school used these "Run, Spot, Run" books, but then she confessed that the children seem honestly to enjoy them. (I had forgotten to ask my son what *he* thought of his reader!) Apparently one of the great anxieties of a child entering first grade is that he is going to be expected to read more difficult books than he can handle. When the child finds the first readers easily manageable, there is an immediate surge of confidence and a sense of accomplishment, which is exactly what the teacher wants. The child is soon into readers that *are* creative and challenging.

Then Mrs. M. handed out to each of us parents a mimeographed sheet containing these cryptic lines:

@#+=   +=   &+$$.

&+$$ +=   %@   @#*   !%().

#*(*   +=   /+$$.

/+$$ +=   %@   @#*   !%().

#*(*   +=   &*>.

&*>   +=   %@   @#*   !%().

+=   “%<   %@   @#*   !%()?

We were puzzled momentarily by the mess of symbols, but then we were told that this was nothing but a simple substitution cipher for “At the Park,” the first story in our children's reader. Mrs. M. had given the little puzzle to us that we might appreciate the obstacles to reading when one is first learning the code of the alphabet. With the adult's twenty or thirty years' experience it is easy to recognize words like *the* and *is,* but to the beginning reader each new word is a new cryptogram to decipher. Even the commonest of words can be troublesome, like *the* or “@#*,” which must be learned by rote, for its spelling is not phonetic. Thus what may be trivial to some is enigmatic to others.

When the children had demonstrated that they had indeed mastered the code of the alphabet, they would advance rapidly through the increasingly difficult readers in the series, we were assured. And the reading book that Mrs. M. held up as the goal of the first grade was more than some of the parents had tackled in many a year.

After our little lesson in humility we were invited to view the exhibit of books on the front desk. There assembled were the collected works of the author/consultant team I had been so critical of just a few nights before. Their task suddenly seemed monumental, and I was grateful for their efforts, flawed as they might seem, to give my son and his classmates accessible reading material for a good year's work.

Although I still had reservations about filled-in *O*s and careless syntax, I left the classroom privately embarrassed over my too-easy (and probably flawed) criticism of those who take the time to think and write texts for our children. At home I told my son that I sat in his chair at school and listened to his teacher tell us what they did at school. I told him that he had a very nice teacher and very good books to read.

When I first reread the opening part of this essay, which I wrote the night before curriculum night, my initial reaction was to scrap it. But being proud even in my humility, I retrieved it from the wastebasket and doubled its length and then appended this penance. May it remind me not to judge so hastily, for tonight my son brought home another, more difficult and less boring reader.

After I wrote the above installments in my continuing Adventures in Kinderland, I thought they made a nice little essay and showed them to my wife, who is my first and fairest reader. Her reaction was quick and to the point: I did not answer in the second part the objections I had raised in the first place. I did not convince her that everything was all right. If indeed my son's reader is as boring and grammatically shoddy as I claim, why am I so willing to praise it to the child?

Why indeed? My wife was right. The little trick of encoding the story proves nothing. It only distracts us from the real point.

My son and his classmates can already read individual words—they do know the code. The longer it takes them to get to "real" books the more risk there may be that books will lose their attraction. After all, through how many "At the Park" stories can they last?

And what about the filled-in letters and the vague references in the first reader? I had no answers for my wife, and I don't think anyone can have any for me. So what happened to me at curriculum night?

I guess I wanted to believe that the teacher and the curriculum and the books were as good as one could expect. If they were that good there would be no need for confrontation with Mrs. M. My experience with my daughter, who is five years ahead of her brother in school, has been that when the parents question the teacher it is often the child who has to answer. If a parent persists, the teacher can play her trump cards—including her treatment of the pupil and her comments on the child's permanent record. So I accepted too eagerly Mrs. M.'s explanation of the series of reading books in order to avoid the very questions I had raised.

Realizing this, I looked again at the second reader my son brought home. Here is its first story, "Who Said 'Hello'?", in full:

> Bill said, "Come with me, Ben.
> I want to see the zoo."
>
> Ben said, "Stop, Bill!
> Who said 'hello'?
> Who said 'hello' to me?"
>
> "We will see," said Bill.
>
> Ben said, "Help me, Mother.
> Who said 'hello'?
> Who said 'hello' to me?"

Mother said, "Can't you guess?
Guess who said 'hello,' Ben."
"I can guess," said Jill.

"A parrot!" said Bill.
"This parrot can say 'hello.' "

"Hello," said the parrot.
"Hello! Hello!"

Ben said, "You can say 'hello'!
Say 'hello,' Parrot,
Say 'hello' to me."

This is what happens when only thirty-eight "new basic words" are added to the fifty-eight maintained from the earlier readers. How much patience would you have reading this insipid dialog with its distracting quotes within quotes? (Does new punctuation not count as words?) The more into the book I read, the more I realized it was more of the same.

Now repetition is a staple ingredient of children's literature, but it is repetition of pattern and tone that works. Repeating in an eighty-five-word story the word *hello* twelve times and using the verb *say* eighteen times, fourteen in the form *said,* does not work. At least it doesn't work for me. It is boring at best, and "Who Said 'Hello'?," with or without illustrations, is downright confusing reading to me.

So I am back where I started, with my first impressions about the school readers. What shall I do? I don't want to criticize the books or the teacher my son has to spend six or seven hours a day with for the next seven months. I'm afraid to risk the consequences of that. I also don't want to argue with a teacher who has thirty-five pupils to contend with. She can't possibly have time to reconsider the outline of her curriculum-in-progress or the selection of books. What do I do about these rotten readers?

Yesterday my son asked me to take him to the library, to the back of the junior room where his sister gets books. Before first grade, by his own choice, he had largely limited his browsing to the low shelves of *E* books, but yesterday he walked once again with me between shelves that were taller than he. He wanted to find books on ponies, cats, jaguars, jokes, and riddles.

I showed him the card catalog, and we looked up his subjects one at a time and found its call number. He caught on immediately, and we selected three books, two on cats (including jaguars) and one on riddles (including jokes).

At home we read the riddle book first, and it had, in the fashion of its kind, the answers printed upside-down below the questions. As my son asked me each riddle, I confessed I didn't know and asked him for the answer. When he read it to me directly from the upside-down code of the alphabet, I knew there was little reason to worry about the school readers. If my son and his classmates progress slowly through the boring primers, they can progress rapidly through the public library.

There the children know the books to select: the soiled ones with the covers tattered and taped. Those are the exciting books scores of children have enjoyed. The boring books are easy to avoid: they are the ones that look like new, like the school readers.

So my son's bookmark progresses slowly through the solemn primer that reads to me like a parody of its genre. But the pile of books beside his bed grows as he adds to it each night. He knows how exciting reading *can* be, and no boring primer is going to slow his progress through the library.

Teachers probably know all this. They probably know that good primers would be devoured in a day, and that no school is rich enough or large enough to buy and store as many books as beginning readers would need. They probably know all this and that is probably why Mrs. M. encouraged us parents to take our children to the library as often as we can.

We do.

18 · Bullish on Baseball Cards

My eight-year-old son began collecting baseball cards this season, and it is none too soon. The activity not only provides him with a link to the sport during a players' strike, but it is an initiation into the world of American business. For the lessons a boy learns from managing a shoebox full of baseball cards are as hard and sobering as any gotten over the bargaining table in a complex labor dispute.

Baseball card collecting teaches its participants all the elements of successful business practice. While it is of course not the only introduction to the realities of the business world, it is a traditional one for boys. To collect and trade cards requires a youngster to become comfortable, or at least proficient, in buying, selling, trading, risking, and generally managing his most important and valuable possessions at the time. There must be an awareness of worth and a deliberateness of purpose in trying to decide whether to specialize in getting the entire home team or to diversify by filling out an all-star team of interleague dimensions.

Assembling the capital of a baseball card collection is as instructive as any other stage of the activity. The boy must make what are usually among his own first purchases that cannot be wholly eaten or drunk. Although the wrapper may be ripped and discarded like that of a candy bar, the baseball cards and not the token bubble gum are the treasure inside. Buying a package of cards is as important to the young boy as purchasing a stock is to his father. But the element of mystery, the joy of the blind gamble of the purchase, is unique to baseball card collectors and venture capitalists. Unwrapping the package is a holiday, and scanning the contents is an adventure.

On the backs of his baseball cards a boy's fascination with numbers begins, preparing him for the Dow Jones industrial average, the Consumer Price Index, and Moody's credit ratings. Besides the heights and weights and birthdates, there are the columns of performance figures that contain as much mathematics and as many subtleties of interpretation as an adult will encounter in reading the market reports.

The young card collector soon learns to scan each new card for his touchstone statistics as naturally as the commuter scans the newspaper to follow his investments. The boy with a baseball card collection knows averages, records, and statistics the way a market analyst knows prices, dividends, and betas.

The seemingly idle hours spent reading the fine print behind the picture of a baseball player are rewarded in eye muscles attuned to the format of the business pages, not to mention in mental skills favorably inclined to mathematical achievement tests.

A boy's trading cards also give him the chronology of a baseball player's record through obscure minor league teams to success in the majors. The repeated reminder that success and celebrity come of humble beginnings is precisely the lesson one needs later to put his foot firmly on the bottom rung of the corporate ladder and begin his climb with deliberation.

Perhaps the greatest lessons of baseball cards are learned on the trading floors of boys' bedrooms. Here hard decisions and value judgments must be made by children alone in a world of cardboard adults. The worth of a card wanted must be weighed against that of a card or cards that might be traded for it. Order must be maintained among piles of cards numbering in the hundreds to the thousands, none of whose provenance or ownership is documented or documentable beyond the agreements of little gentlemen.

My son is now investing virtually all of his modest allowance in baseball cards, and he is bullish on new issues. His present portfolio consists of about 100 Fleer, 100 Don Russ, 200 Topps common, and an odd lot of autographed Topps preferred. He holds his own trading with the older boys, and soon I expect him to be willing to trade more than just the Yankees for a share of Telephone.

Not all traders are gentlemen, of course, and stories are told among the eight- and nine-year-olds of eleven- and twelve-year-olds to be watched. And the younger traders learn quickly that it is against their interests to give in to unreasonable demands for a card they want, no matter how badly. Even a Fernando Valenzuela is not worth the whole Yankee team.

For all of its traditional values, however, there is one aspect of business that baseball card trading does misrepresent. It is a very sexist thing, and there are no women on the cards. Thus the hobby is practiced among a network of good young boys who may grow up thinking the good-old-boy network still excludes women. It does not, of course, but after their team and position, the hirsute heroes on baseball cards are described by the vital statistics of height and weight—6 feet 3 inches, 190 pounds—which all the boys want to attain. There are no models here for girls.

19 · Outlets for Everyone

My son, who is ten years old, covets everything electronic. His latest desires include a home computer, a stereo, a miniature television set (including a subscription to cable TV), and a video cassette recorder—all for his own small room. These electric appliances would join the telephone, clock radio, and numerous battery-operated electronic games that he already has scattered among the softer and quieter objects of a generation ago. The balls for all seasons, the various shoes that go with the balls, the variegated uniforms that seldom match the shoes, and the general insignia and ephemera of sport that remind him (and me) of games past, cover the floor in a deep cushion of plastic, leather, rubber, and foam, as if to provide a safe surface over which to cradle the latest generation of electronic devices that should not be dropped.

For as long as he has been old enough to watch "Sesame Street" and "Mister Rogers" on television, my son's room has emitted the sounds of an electronic jungle full of beasts. Their multitudinous and eerie red eyes peer out at him from the dark beneath his covers long after he has been told lights out and has put down not the Sears catalog but the latest catalogs of discount chains, the preeminent purveyors of electronic merchandise hereabouts.

In his covert tent, like generations of children before him, he is busy reading—but not my books, nor his grandfather's. And he needs no flashlight. Instead, he reads the keyboards of the toys of a new age as a blind man reads Braille. And here, after lights-out, he continues to learn about computers and microprocessors—the innards of his plastic friends—as he learned math with the Little Professor and spelling with Speak & Spell, by playing. He mastered the rules of hand-held baseball and basketball games long before he learned the rules of the real games. Tonight he plays a hand-held version of Donkey Kong. Outside the world of brave Mario trying to rescue the maiden Pauline from the mighty if dopey Kong, dead batteries lie scattered about the bed like little lead soldiers, and the wall outlets buzz with the mysterious energy recharging the rechargeable.

I never pressed the buttons of a keyboard until my mid-teens, when I learned to touch-type on a sluggish mechanical device that had its letters masked so that I could not peek at them even with a bright flashlight. And I did not know the feel of an electronic keyboard until I exchanged my slide rule for a hand calculator. Thirty years ago, when I was ten years old, the night jungle beneath the covers of my bed was lit by a forever-dimming white-eyed cyclops whose batteries were not rechargeable, and my ears were not shattered by the simulated screams of maidens, the stomping of carpenters, or the roars of gorillas. My fingers turned the pages of *Treasure Island* and *The Three Musketeers* as they would later turn the pages of Wallace Stevens:

The house was quiet and the world was calm.
The reader became the book; and summer night

Was like the conscious being of the book.
The house was quiet and the world was calm.

The sounds I knew in the night were the quiet and reassuring ones of the rustle of the leaves of books, the ticking of a hand-wound alarm, the distant whir of the electric kitchen clock, and the intermittent hum of the refrigerator. These were the precursors of the ceaseless sounds of ubiquitous electric motors that Norman Mailer heard in the distance of time.

Only when I became a full-fledged teenager and coveted 45s to play on my portable record player, a transistor to take to the beach, and any old AM radio to listen to as I lay in bed, can I remember tapping the electricity from a battery or a wall receptacle. Now the outlet is the inlet to a world of home entertainment —not only private, under-covers games, but also the video games and tapes played before the electronic hearth of the late twentieth century.

I have often thought that the social history of electricity could be documented by a chronological series of drawings or photographs of the wall outlets that brought the fickle fluid from the dynamos of the nineteenth century—whose power Henry Adams felt charging the atmosphere of the future—into the homes of the twentieth.

In the hardware store/museum of my mind, all wall outlets look like totems. They chronicle the changing lifestyle that came with the electrification of this country. The holes that receive the prongs of plugs are the eyes, and the general overall shape of the receptacle, surrounded by the cosmetic wallplate hiding the increasing wrinkles of wires, is the face. In the images I retain of the earliest electrical outlets, the eyes are round, representing both the enchantment and the apprehension of our grandparents toward their new technology. As their early fears were allayed, our ancestors and their wall receptacles grew older-looking together, their eyes becoming lengthened and sunken with age. The wide-eyed days were over, and electricity was ho-hum stuff. Outlets became stoic in their stare, and they were stacked one on top of another as the figures on totem poles and apartment dwellers and commuters are wont to be, and the round, youthful shapes of their faces gave way to those neither round nor square truncated pumpkins shriveling forgotten in the chilly autumn after Halloween.

Now, rejuvenated via the life of condominiums, microwave ovens, personal computers, and electronic games and appliances of all manner and kind, the older generations and their wall totems have come to look animated again, with open mouths wondering what third prong of what plug will be inserted next. When it does not look like a robot, grounded in all the latest technology, the modern wall outlet looks almost human, especially when we notice the essence of humanity, the asymmetry of the eyes, designed to receive the asymmetrical prongs of polarized plugs to protect the devices at the other end of the cord.

Plugs and outlets are seldom the focus of the electronic revolution, however. They are merely the connection it takes a James Burke to recognize and explicate. What we tend to think in terms of are the termini of the connection. Behind the wallpaper scenes so cleverly matched on many a cover plate is the source of electricity, somewhere over the rainbow, perhaps symbolized in the arch of a hydroelectric dam or the hyperboloid of revolution of a nuclear reactor's cooling tower. On the other end of the long extension cord is the appliance that today could be anything from a child's toy to a personal computer to an artificial heart.

If older generations, mimicking the outlets of their own youth, are awed by the universes captured on silicon chips of sand, the younger ones seem to take them for granted and welcome each new wave of them with about as much marvel as they would another high tide at the beach. My son's generation has come to expect new electronic games and gadgets as I did new books and records. And as my eyes and hands needed no time at all to adjust to a new style of binding on an open book or a new typeface on a spinning record label, so my son and his friends grasp each new electronic toy with confidence and with what seems to be an innate understanding of its function and the rules of the game. In fact, I understand that children do not learn how to play a new video arcade game by reading the instructions. They just drop their quarters into the slot and get the feel of the joystick or whatever else controls the motion of the turtle or dot-eater, and they learn by what is nowadays called hands-on experience. That's a far cry from the reluctance of some of my parents' generation to even fly in an airplane.

There has been considerable talk about what has come to be termed computer literacy. The argument is that the computer will be an integral part of everyday life in the not-too-distant future and that those who do not speak its language will be as disadvantaged as those who do not read or speak English well enough to be considered functionally literate. Watching my son and his generation, who seem to pick up the language of computers by the same mysterious assimilative process whereby they pick up the ordinary language of their parents, I have little fear of any real computer illiteracy in his and succeeding generations. While it may have been the case that previous generations could ignore such basic technological skills as driving a car, the new generations cannot do the same with computer skills. But the electronic game revolution has made computer literacy part of play, and no child does not know how to play. And as the prices of the devices made of the chips of the stuff of sand continue to drop as they have been doing, the essence of a computer will be as available as a bottle of soda or a rubber ball.

For the more fortunate among us, the toys of technology have even become the essential luxuries. My son has his video-game computer attached to our family television set, and each new game cartridge that becomes available is as important for him to acquire as books are to his parents. He pleads with us for the game to top all games, and we eventually give in for reasons that alternate between weakness and pity (on the record; off the record we are curious to see what they will think of next).

When the new cartridge is brought home and inserted in the master console, our son becomes glued to the television and the connected joystick until he has mastered the game. Mastering the game does not necessarily mean getting a perfect score or even winning: it soon becomes clear to everyone that the computer always has the advantage of the house. Rather, mastering the game is understanding the computer game's logic to the point where its advantage is accepted as one accepts the law of gravity, the law of diminishing returns, and the law of averages. The game's the thing, for it can't be whether you win or lose.

Often my son and his friends will begin to play *with* the computer rather than against it. That is, they show how well they understand the workings, random as they may be, of a particular computer strategy. My son has demonstrated to me this mastery of the computer by showing me how he can score just as high with his eyes closed as with them wide open. Or he will play with his back to the screen, hitting bull's-eyes as well as Annie Oakley. He has shown me how he has caught onto the computer's inner clock, which is, overall, very predictable for all its local randomness. When this stage of mastery is reached—often after only a few days with a new video game cartridge—the rules of the game are no longer under the control of the computer. Now my son wants to get his own computer so that he might program games that he devises himself, thus not only controlling his part of the game, but the computer's as well.

Just as our children come to us one day to show us how they can read the book we have read to them at bedtime for so long, so they will come to us and show us how they can read the minds of computers they have played with for so long. This is why computer literacy—though it may be a temporary problem for the generations who have had computers thrust under their fingers after they were ten years old—should be no real problem in the long run.

My fifteen-year-old daughter is a child of the transition. She learned to read before "Sesame Street" and Speak & Spell but did not reach adulthood before computer games. (That is how quickly the electronic revolution has occurred.) Her childhood includes memories that do not require batteries or electrical outlets, and yet she need not assume a pseudonym to talk about the pleasures of playing Ms. Pac Man (to use as an example the video arcade game whose sexual liberation is so curiously confused and incomplete). Even though my daughter did not need "Sesame Street" to learn the alphabet, she watched the show from its premier. Even though she did not need arcade games to learn about LED displays, she has played her brother's games from the first. But, unlike him, she can take them or leave them.

Just as my daughter was born in medias res, so her room—which is in the part of our house built before the Beatles sang—was constructed before electrical codes called for the plethora of outlets that they require today. As a result, her room is electrically deficient for her present needs. While she has not coveted the same things her brother does, she has not been the least bit shy about using things electric. She has—in addition to the common lamps and a clock radio—a stereo, a television set, and numerous electrically driven beauty aids. Her hair has been dried and curled and styled with more cleverly named and designed appliances than she can find outlets for. She has had to resort to using extension cords to multiply the sources of electricity in her room, and still she often needs to unplug one device before using another. In the evening she places her contact lenses in an electrically-operated bath that cleanses them for another day, but only after she has arranged the plugs that she will nearsightedly need to find in the morning.

For those of us who have not grown up with electric beauty aids and lens cookers—much less computers, magic calculators, and electronic games—the fear of computer illiteracy can be a self-fulfilling prophesy as real as the fear of flying. Since my wife and I are both writers, the big question in our household several years ago was whether or not we should get a word processor. I generally argued that we did not need one, for we had gotten along fine with the typewriters we had learned to touch-type on long before either of us knew what a computer even was. Since I was writing mostly short essays and poems then, the ordeal of retyping a whole manuscript did not ever cast a cloud over my future. However, my wife was writing a novel, and retyping four or five hundred pages of manuscript every time she made a revision was not her idea of fun in the electronic age. She began to look into word processors as assiduously as my son does video games. She systematically studied the possibilities and compared the specifications in search of a good basic word processor—and she got one. The necessity of revision was the mother of intention.

Unlike my son, however, my wife had no background in electronic games, and she knew nothing about operating computers when she began using the one she bought. But, as my son knew he wanted to play the games computers could be programed to play, so my wife knew she wanted to enjoy the convenience she knew a word processor could provide. She knew that once she typed a manuscript into a word processor she could revise it at will without ever having to retype any parts except those she wanted to change. She could add, delete, and rearrange whole blocks of words, sentences, and paragraphs at the touch of a key or two, and in the end she could have a copy printed out that would be virtually free of typographical errors. It all sounded too good to pass up.

My wife never worried about the fact that she knew not the first thing about computers. Since she could read a cookbook to plan a dinner or a textbook to prepare a lecture, she figured rightly that she certainly should be able to read a computer manual to process its words and hers. And so she did. For a week, she studied the manual night and day and finally got the word processor to do what it was supposed to do. I generally stayed out of her way during that week. My patience is not so great as hers, and I played video-game tournaments (with barely an audible beep) with my son, who always won. When she did emerge from her retreat, my wife showed me computer games that made writing more fun, even for someone who dealt only in shorter manuscripts. And these were computer games of solitaire that I could win.

Since my initiation I have become a convert to the machine with the memory, and I am writing this true confession directly into it. Last summer I typed my first book-length manuscript of prose onto its floppy disks, and I believe I found the stamina to do so only because I knew that I might never have to retype the whole thing again. As the novelty of the word processor has worn off, we all have become as accustomed to it as my son has always been with his little plastic boxes of buttons and lights. My daughter has used it for her school reports, and she and my son are beginning to program games onto disks of their own.

The computer has changed our work habits somewhat, and it has necessitated our moving the furniture around a bit in our study. Where there was a single typewriter before, now there is a keyboard, two disks drives, a display terminal, and a printer. These objects take up three electrical outlets, and there are not many left in the study that only a year ago had too many, or so we thought. But who is to say that our electric pencil sharpener is a luxury in the computer age, or our telephone answering machine, or our recharging tape recorder?

The plugs of these appliances fill the happy receptacles in this room we built only two years ago, when we knew we could not work in the same place with the television, the stereo, and the children watching and listening to them and to us. When the electrician told us that he had to put in so many outlets because the building code required them, we were a bit skeptical and thought how convenient for the electrical contractors. We remained skeptical for a year while only a lamp here and there and an electric typewriter and an odd pencil sharpener or recording device required a source of electricity. But now we are using virtually every one of the twelve outlets in the room of modest proportions, and we wonder if we will find ourselves holding a homeless plug one of these days.

It is sometimes fashionable for those of us who grew up without so many conveniences to ridicule the dependence of our dependents upon them and to lament the passing of the days of self-reliance and roughing it with little more than an electric fan. But it is not so easy for the younger generation. The products of technology evolve so rapidly that our children see their toys and necessities becoming obsolete before they tire of them. Children today are experiencing in their toys what we called planned obsolescence in the era of chrome trim and grilles and fins on automobiles, when each model year brought a new look in cars as unpredictable and arbitrary as the hemlines of skirts and the number of buttons on sports jackets.

When I was a child I put away each year the appurtenances of one season's play for the next. A football was a football and not a Nerf football or some other annual variation, and electric trains were something we added to each year rather than replaced with the latest slot racers, which seem to be as disposable as the razors they resemble.

Children no longer can feel secure putting away the toys of this season for the next, however, for in their few years they have learned that what is in this year will most likely be out the next. Thus they have come to read the catalogs of discount stores as if they were fashion magazines. Almost imperceptible differences among the pages of electronic games, appliances, and living aids become the talk of the playground and the park, and children have come to await the new catalogs the way their parents did the Sears catalogs of yesteryear.

Our children's eyes have become accustomed to reading the

small print of television commercials—the print that tells that batteries are not included—the way I learned to read the label on a 45-rpm spinning on my friend's record player. For all its visual pyrotechnics, the electronic revolution has not reduced the need for children to read. If anything it has made reading acuity even more important for a consumer's survival among the electronic vines of a micro-jungle.

Though they don't need manuals to learn the rules of a game, our children know they need catalogs to learn the games that rule. The magazines of the electronic age are beginning to appear in great numbers, and the video gamers are reading them to get the latest rundown on what is de rigueur this month and to find out what are the most recent high scores to beat. My son waits for his *Odyssey Adventure* the way children of my age waited for *Model Railroader* to come in the mail.

The current plethora of electronic things is but the subject of another chapter in the history of technology and the world. Just as *plastics* was the buzz word of the 1960s, so *microchips* is that of the 1980s. I found plastics and everything that the word implied then abhorrent until I went to Mexico and saw the markets filled with the brightly-colored plastic products that made the mundane chores of life a little bit more convenient for poor Mexicans. Now the microchip can do the same, and children should be allowed to enjoy the electronic game that will soon be as relatively inexpensive and ubiquitous as the hoop and stick of another time.

It is as easy to say that the products of microelectronics are unnecessary as it was to say a century ago that electricity itself was unnecessary. But what is a frivolous luxury today can be a commonplace tomorrow. What is one generation's toy is another's laboratory apparatus. Our children are growing up in a world of electronics that makes "The Milton Berle Show" seem like a French cave painting.

When I was young I remember fuses blowing all the time. I

remember my mother having to remember to turn off the toaster before vacuuming the rug. I remember my father putting a coin behind the fuse to keep it from blowing every time the new refrigerator started up. Now I live in a house that has circuit breakers instead of fuses, and I can remember the circuits breaking only one or two times in the past three years. For all our appliances operating full tilt with one another and for all our abandon in operating everything with everything else, our children do not seem to know the meaning of "to blow a fuse." To them, the term is but an obscure idiom of the English language having something to do with getting angry. And they can remember no power failures. Those that have occurred have done so only during the hours of sleep, and only the blinking digital clock gave any hint that the flow of electricity had been interrupted during the night's storm.

The delivery and use of electricity has become so dependable and commonplace that, to our children, it waits in the outlets the way water waits in the faucets for our command. All that remains for our children to do is to pick the correct plug from among the tangle on the floor of their rooms and to find the face of a ready and willing outlet. To their parents it all may be a miracle, but to the children, it is as natural as whole wheat bread.

Our family never sits around a radio listening speechless to the magic of it, as I once did with my parents. The radio is a given of the household, hardly noticed. So, too, will be the new electronic instruments of home entertainment. It is as surely a part of nature for the transistor to have evolved into the microchip as it was for radio to have evolved into television. What will happen next in home entertainment and home appliances is as open to speculation as what will happen next to man.

20 · Is There a Big Brother?

MCMLXXXIII

Even as George Orwell was writing *1984*, his futuristic novel warning about the control of society through technology, the technology was already being developed that could fulfill—and even exceed—some parts of Orwell's nightmare, while at the same time forestalling—or even making impossible—other parts.

The first all-electronic digital computer was completed only four decades ago, in 1945. Back then, one cumbersome machine's myriad switches and mazes of connections filled a whole room at the University of Pennsylvania; today, hardware with the same computing capabilities is portable and can fit easily into an attaché case. This degree of miniaturization is only one aspect of computers that was not easily foreseen in the late 1940s. Another is the ability of today's computer to store and manipulate information, an aspect that is wide open to abuses that go far beyond what even Orwell imagined. While it is unlikely that social control will ever actually reach the degree of totalitarianism envisioned in *1984*, today's computer technology already has the potential for making one tool for that control—the rewriting of history—even easier than Orwell depicted.

In *1984*, protagonist Winston Smith works at the Ministry of Truth. His job is to rewrite stories in old issues of the London *Times* in order to make speeches by party leader Big Brother appear to predict correctly what subsequently did happen. Smith reads through archival issues of the *Times* in which Big Brother did not demonstrate omniscience, then dictates new versions of speeches and stories to rectify the flaws. Smith also corrects other factual errors, changing the name of the current "enemy" of the party or eliminating all references to people who have become "unpersons." New editions of the paper are then printed in the conventional way to replace the "incorrect" original ones, which are tracked down and destroyed by other employees of the Ministry of Truth. Today technological advances not only enable a Winston Smith to do his job at a computer terminal in his own home instead of having to commute to and from the ministry each day, but they could also make it unnecessary to track down and correct every existing copy of the supposedly errant newspaper.

Electronic versions of such newspapers as *The New York Times* and *The Washington Post* are already available to owners of home computers. Instead of looking for the morning paper outside his door, the modern subscriber can connect his computer terminal through his telephone to the newspaper file. He can then scan the headlines, summon selected stories, and electronically command the computer to search for stories containing key words or phrases. The same newspaper stories are written by reporters at other computer terminals that have above their electronic keyboard a television screen instead of a paper carriage. Of course, hard copies of the newspaper are still also printed and distributed in the conventional way, but it is conceivable that, someday, newspapers will only exist in computer memories and be read at computer terminals.

Typesetting is already done electronically by many of the largest newspapers, including the London *Times* that Orwell saw as the final arbiter used to document official history in *1984*, and it is a natural step to store past issues electronically rather than in cumbersome piles of paper that take up space, turn brown, become brittle, and tear. When I recently tried to get a copy of a 1981 edition of the *Kansas City Star* I could find no actual copy of the paper anywhere. All libraries referred me to their microfilm files, which they said took up much less space than the bulky old newspapers that few people ever read. Even the *Star* itself sent me a copy made from its own microfilm files. The same rationale makes a central computer file, accessible by telephone by libraries and individuals everywhere, so attractive. No library—or newspaper—would ever have to worry about space to keep past or future editions of the newspaper on microfilm. If newspaper files were to exist entirely in the memories of computers, then everyone's "copy" could be changed by changing the memory. Only the eccentric who might make personal copies of the electronic newspaper—either on printout paper or in his own computer memory devices—would be able to check his human memory of what really happened back when.

For some years now, afternoon dailies have been finding it more and more difficult to compete with the evening news on television, which ultraconservatives have expressed considerable interest in controlling. The *Washington Star* and the *Philadelphia Bulletin* are among the not-too-remote casualties in the war between print and video journalism. As younger generations, who have learned to talk and read in front of television sets, grow into consumers of news, sports, and weather reports, even the morning newspapers may find themselves threatened with diminishing circulations. Publishers may naturally encourage the selling of subscriptions delivered to computer terminals; computer marketers may in turn feature the new mode of newspaper reading in their ads to sell computers.

Since it is likely that a great many people will be working at home at personal computer terminals instead of traveling to and from work each day, the demand from commuters for hard copies of newspapers could diminish to insufficiency. Electronic subscription to the *Times* of London, New York, or the world could be the only way to get the news; the presses could be stopped forever. With electronic typesetting and delivery, no paper need enter a reporter's typewriter and none need exit a printer's press. The electronic circuit could close upon itself, caging the truth behind magnetic bars that can dissolve beneath the keeper's touch. Because the "truth" will be so easy to change, as in *1984*, it could be very difficult to know. By the end of this century, the three great electronic inventions of the last one hundred years—the telephone, the television, and the digital computer—could be linked into a paperless information network that could indeed have the potential for changing not only the future but also the past, which might be stored almost exclusively in an electronic memory.

There already exist in England, France, Canada, and the United States prototype systems of information transfer, and the future of such systems is a trillion-dollar market worldwide. The ongoing wiring of the United States for two-way cable television is startlingly reminiscent of the "telescreens" in *1984*—the electronic receivers that doubled as transmitters, allowing Big Brother not only to broadcast propaganda but also to eavesdrop on every citizen's domestic activities.

Shared data banks, interlocking electronic directories, and secret files revealed only by the pranks of teenage beeping toms, are already among the horror stories of our new computer age. Yet no matter how real these fears may be, there is also reason to believe that the computer revolution has progressed so quickly that its technology has already evolved to a point that would give Big Brother himself fits and thwart his every effort at mind control.

In *1984* Winston Smith kept his secret diary in an old-fashioned copy book, where he wrote (just out of sight of the telescreen) unthinkable thoughts with an old-fashioned pen. Had Big Brother's informer burst into Smith's apartment while he was making an entry, the secret would have been out—a whole copy book cannot be swallowed as quickly as a floppy disk can be erased. And it is the *fear* of having one's privacy invaded, manifested in any high-tech snooping device, that alone could enable a Big Brother to oppress a society.

Had Orwell been a true visionary, he might have given Smith a word processor, the copy book of the real 1980s, to go along with his computer terminal. Even before 1984 we knew the cathode-ray tube of a powerful desk computer to be the medium of the creative and introspective as well as the journalistic word, presenting to the upright erect writer a virtually bottomless page. Or so it seems when one first plugs in his new hardware and calls up his software. But soon the light writer learns that a floppy disk has only so much space—not much more than an old copy book. And as we learn the computer's limitations, we begin also to grasp that data banks, directories, secret files, and central newspaper archives also cannot be maintained or added to without limit.

Thus the computer revolution, rather than hastening the dawn of *1984*, should ultimately prevent it. As more and more of us learn not only the convenience of the computer but also its limitations, we learn to distrust the horror stories of central government computers recording our every move or our every transaction, and we learn to look with suspicion upon any newspaper archives that are not backed up with hard-copy evidence. Knowledge brings forth a liberating skepticism where technological ignorance can cast an oppressing spell of omniscience. Technological literacy enables one to distinguish engineering fact from science fiction.

Rather than bringing Big Brother bullyingly closer, then, the computer might in fact be the very device that will hold him at bay. Real children of the real 1984 grew up with computers. And by growing up with them the computer generation will know the computer's weaknesses as well as its strengths. The machine will not be able to bluff them. Already the microcomputer has become as much a part of many homes as the television, but families will not be cowering passively before the green screen the way fictional characters did before the telescreen in *1984*. The computer terminal will be a friend who is there when needed, whether to be an opponent in a video game or an amanuensis when a report is due at school or at work. Benign interaction and cooperation among people and these machines is what the future holds.

Savvy children of the electronic age know that no computer network can watch all of the people all of the time. They also have already proved that their stand-alone microcomputers can more easily monitor than be monitored. Thus it is the megalomaniac who would be Big Brother that must watch what he writes in his electronic diary or data bank in the mid-1980s and beyond. The millions of computer-literate children out there will have *his* number (or find it by hook or by crook, by trial and error) and will be keeping an eye not on the Winston Smiths but on *him*, whether he be individual or bureaucracy. And since he will be outnumbered millions to one, any aspiring Big Brother will have not a minute or a secret to himself. That was the true meaning of children everywhere asking for computers for Christmas in 1983. They were singing, "Joy to the microworld! Happy 1984!"

But some of their aunts and uncles, their parents, and their older siblings, like George Orwell, children of an age without computers, still believe in Big Brother and warn their children of his existence as surely as those skeptics of almost a century ago *denied* the existence of Santa Claus. And just as a little girl was moved to write to the *New York Sun* in 1897 asking if there was a Santa Claus, so it is not hard to imagine that her namesake was driven, in the waning days of 1983, to her personal computer to compose and print out a letter to the editor of the *Sun*'s successor. Her letter and the editor's response—no doubt the sane voice of Christmas past, present, and to come—would speak for themselves, perhaps in a tone very close to the "Yes, Virginia" editorial of a computerless era:

Dear Editor:
Some of my older friends say there is a Big Brother. Daddy says, if you see it in the Sun, *it's so. Please tell me the truth, is there a Big Brother?*
Virginia Smith, 1984 Prospect Avenue

Virginia, your older friends are wrong. They have been affected by the fear of a fearful age. They believe everything that they see on the television or computer screen. They think anything must be real that can be stored on a cassette tape or a floppy disk. But all memories, Virginia, whether they be men's or children's or computers', are finite. In this great universe of ours Big Brother is a mere dozen or so keystrokes of input, compared to an unprogrammable intelligence capable of grasping the whole system of truth and knowledge. A computer is a mere grain of sand, a silicon chip, in its memory, compared to an intelligence capable of thinking up the silicon chip, or a poem, or a symphony.

No, Virginia, there is no Big Brother. He does not exist as certainly as love and freedom and discovery do *exist, and you know that they abound and give to your life its highest beauty and joy. Alas! How dreary would be the world if there were a Big Brother! It would be as dreary as if there were* no *Virginia. There would be no childlike faith then, no imaginary friends, no Trivial Pursuit to make tolerable this existence. We would have no enjoyment except in sleep and a blank screen. The eternal energy with which childhood fills the world would be consumed.*

Believe in Big Brother? You might as well believe in ghosts! You and your friends might tape every television channel all night long on New Year's Eve to catch Big Brother, but even if you saw him in a video, what would that prove? Everybody can see Big Brother everywhere! The most unreal things in the world are those things that children and men can see on screens. Did you ever see ghosts dancing on the television screen, or snow in the middle of summer? Of course, but that's no proof that they are real. Anyone can see every terror imaginable and conceivable in the world of electronics.

You take apart your digital watch and see what makes it work, but the real meanings of time and space and life remain the puzzles you will enjoy trying to solve again and again. There is no access code to the mind or heart of even the littlest child. But everyone can break the code of a computer. Faith, luck, fancy, serendipity, and perseverance provide the passwords that access the files of diabolical games and oppressive data banks. And are they real? Ah, Virginia, in all this world there is nothing less real and less abiding.

A Big Brother! Thank God he does not exist, nor will he ever. A thousand years from now, Virginia, nay, ten times ten thousand years from now, he will not oppress the spirit of childhood.

At Play

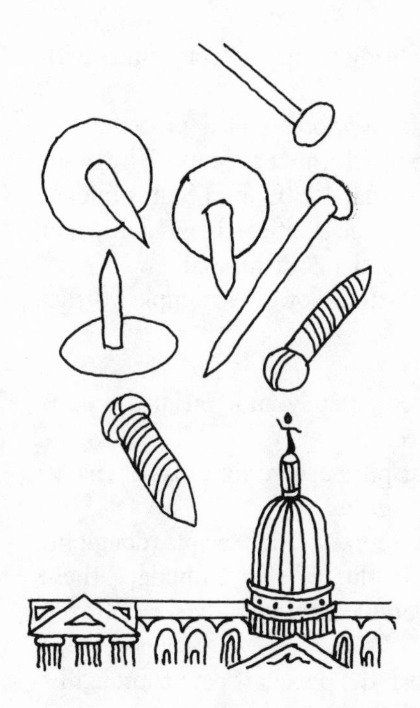

21 · How to Balance a Budget

The 1981 model United States budget was a 636-page tax guzzler that began rolling off the assembly lines of the Government Printing Office in late January 1980. Within weeks, the document was discovered to contain serious design defects, and a massive recall program was initiated.

Potential hazards were associated with a critical part known as the bottom line, which possessed a negative value, called a deficit, thus rendering the whole budget unbalanced. This defect could cause adverse economic accelerators to jam at full throttle and hurtle the whole economy full speed ahead into situations of runaway inflation.

Owners of unbalanced budgets are encouraged to take them to a nearby Federal Reserve Bank for corrective maintenance. Budget cuts and other simple adjustments will be made and errata will be pasted into the documents to bring them into conformity with economic safety standards.

For those who eschew errata, a more radical procedure to remove the budget deficit can be performed at home. The do-it-yourselfer is cautioned, however, that the United States budget is a precision instrument, and the careless citizen may suffer paper cuts and eyestrain attempting to balance it himself.

But if he insists, the individual can follow these simple instructions:

1. Disassemble the budget completely and spread the parts out on a true level surface.
2. Group the parts into neat piles according to type: charts, tables, graphs, numbers, words.
3. Locate the two pie charts, representing receipts (dough in) and outlays (dough out). In the unbalanced budget, these most important parts resemble very unfairly sliced frozen pizzas with the works.
4. Discard the pizzas and wash the pie-chart pans thoroughly. Then set them aside to dry.
5. Using only enough words to make loose connections between bar graphs, columns of figures, and other numerical elements, assemble them all into a structure resembling a roof truss or government-organization chart. This assembly is known as the bureaucratic structure and provides the main supporting arms of a balanced budget.
6. Forget any unused words or numbers that did not fit into the bureaucratic structure.
7. Select a suitable fulcrum about which the bureaucratic structure can pivot. Any symbol of broad-based popular support, such as a souvenir model of the White House or the Capitol, is a good choice.
8. Carefully balance the bureaucratic structure on the White House flag pole, Capitol dome, or fulcrum of your choice.

9. Oil the fulcrum joint so that the bureaucratic structure will be disturbed by the slightest budgetary imbalance or will respond to the gentlest winds of change.
10. Prepare the critical pie charts by spreading equal amounts of dough out on lightly oiled pizza pans. Since the model FY1981 is an austere budget, no pizza sauce or topping should be used. Special decorator fasteners are to be used to complete the pizza-pie charts and attach them to the bureaucratic structure.
11. Open the budget-recall balancing kit (available at Federal Reserve banks) and familiarize yourself with the contents: assorted screws, income tacks, Social Security tacks, corporate tacks, excise tacks, etc. There is also a large supply of windfall-profits tacks, but these are to be set aside for future use.
12. Fasten the receipt pie-chart pan on the left end of the bureaucratic balance arm with tacks. Use one tack of the appropriate kind for every billion dollars of revenue. If you are accurate and neat, your finished receipt pizza will be topped with about six hundred tack heads arranged in sectors whose sizes are proportional to the revenue they represent. The biggest slice of pizza will be covered with more than 250 personal income tacks. It is a hard bite to swallow.
13. Now balance the budget by fastening the outlay pie-chart pan to the right end of the bureaucratic arm with the assorted screws. Use one helmet-head screw for each billion dollars designated for defense, one Phillips-head screw for each billion dollars going to individuals named Phillip, and so forth. Soon the outlay pie chart will begin to resemble a green pepper-pepperoni-anchovy-onion-sausage-mushroom pizza that nobody ordered.
14. When the screws and tacks exert equal pressures on the bureaucratic structure, it will assume a true horizontal posi-

tion. Then your budget is balanced, and you can admire its subtle movements as the internal bureaucracy in the structure hums away and shuffles paper back and forth through loopholes and between the pizza-pie charts.

15. If your budget should become unbalanced during the fiscal year, you may be using too much oil. In that case, return to Step 1 above.

22 · Washington Entropy: Losses from the Energy Bill

The National Energy Act was finally passed in the closing days of the Ninety-fifth Congress, after almost a year and a half of deliberation. Some believe that this comprehensive legislation might have come into being with much less debate had it not been for numerous proposals that congressional subcommittees considered but later rejected as too radical and disruptive of our way of life. Among these proposals, which might have made the energy bill physically impossible to pass, were the following:

Amend the Law of Gravity

It was proposed that beginning October 1, 1980, falling bodies should no longer accelerate at thirty-two feet per second per second. Under this proposal the legal rate would have been lowered by increments until a new rate of twenty-two feet per second per second was established. This would have meant that matter would be attracted to earth with about two-thirds the present pull of gravity, thus requiring less energy for us to climb hills, pitch hay, pump water, etc. New high-jump records were expected to distract the population during the period of adjustment, until it was realized that slower-falling baseballs, being easier to catch, might ruin the national pastime.

Liberalize Conservation Laws

The present Law of Conservation of Energy—the First Law of Thermodynamics—places strict requirements on energy accounting procedures. Another unsuccessful proposal would have relaxed these requirements so that energy systems could be operated on a deficit basis. Power plants would have been able to operate at one-thousand-percent efficiency, and there would have been free lunches for everyone. But this might have touched off food shortages, opponents countered, since we all tend to eat much more than necessary when we are not picking up the tab. The proposal was put on a back burner.

Abolish Entropy

This measure repealing the Second Law of Thermodynamics—the law that requires that the entropy, or unavailable energy of the universe, always increase—would have made immediately available vast amounts of energy tied up in heretofore irreversible thermodynamical systems. Since increasing entropy is associated with disorder, a nontechnical spin-off of this legislative action was expected to be a reversal of society's decline. The proposal became known as the natural law and order provision, but it failed to gain support when physical states' rights legislators realized that it would have denied the inalienable right of all atoms to pursue their own free paths.

Lower the Boiling Point of Water

If, instead of the 212° Fahrenheit of present physical law, steam could be generated at 150°, as required by another provision of the original energy bill, it would have taken less energy to run the turbines in power plants. The proponents argued also that coffee would have perked in less time, allowing commuters to sleep a bit longer in the morning, thus reducing their predawn electricity consumption. While energy experts questioned the technical merits of the arguments, this proposal was ultimately defeated on the advice of a small group of congressmen whose cars are prone to overheat even at the present legal boiling point.

Outlaw Rolling Friction

This was one of several measures intended to reduce the gasoline consumption of wheeled vehicles. A joint House-Senate conference committee considered outlawing all friction until it was pointed out that such a move would have made stopping rather difficult, thus reducing revenue on toll roads and bridges. The whole proposal was abandoned when it was realized that without rolling friction it would no longer be possible to put English on bowling or billiard balls or roll tobacco in cigarette papers.

Develop a Breeder Diesel

If this proposal had found its way into the National Energy Act, a long-range, multibillion-dollar research and development program would have led to an internal-combustion engine that, in its operation, would have produced more fuel than it consumed. When this technological milestone had been achieved, passenger cars would cease to be sold, and only empty fuel trucks would be marketed. This new breed of vehicle would stop at service stations to empty full tanks of fuel into pumps still required to supply conventional vehicles. The provision was deleted when it was realized that eventually the country would have run out of fuel storage facilities and would not have been able to export diesel fuel to an oversupplied world.

Deregulate Time

Daylight Savings Time was to be abandoned under this provision, and clocks would have been required to run faster during hours of peak power demand and slower during periods of low power demand. By this move Congress hoped to achieve an overall reduction in U.S. energy needs. When a young page, who was studying engineering during congressional recesses, pointed out that energy is the product of power and time, thus raising a question about the efficacy of the provision, the whole idea was referred to committee, and it died of neglect since there was no time to consider it.

Postpone Going Metric

Under this provision the present system of measurement would have been maintained until a three-foot meter could be developed. This would have reduced nationwide the commuting distance by almost ten percent. There were also efforts to require a one-quart liter, which would be five percent smaller than the present measure, and a two-pound kilogram, which would be a full ten percent lighter than the current unit. The proposal was shelved when it was feared that inflation would escalate when suppliers charged twenty percent more for commodities sold in the smaller meter, liter, and kilogram quantities.

Provide a Tax Credit for Sleep

For every hour beyond eight that an adult wage earner slept, he was to have received a credit on his federal income tax. The tax credit would have been a graduated one, being greater for light sleepers and less for heavy ones. "Sleep is darkness," one proponent explained, "and darkness keeps the lights out." This provision was deleted from the final bill when the question of electric blankets arose.

Admit the OPEC Nations to Statehood

This move would have immediately reduced our dependence on foreign oil and made the U.S. energy-independent for the foreseeable future. There appears to be no good reason why this proposal was not included in the National Energy Act, for it does not even seem to be prohibited by the present laws of nature.

Impose Import Quotas on Weather

This measure was expected to reduce the number of winter cold fronts, all of which originate outside the continental United States. When Canada threatened to impose its own import quotas on water flowing over Niagara Falls, the proposal was withdrawn.

Create a Department of Lethargy

This new cabinet-level department would have encouraged inactivity, thus reducing per capita energy consumption. The department was to be known as DOL, which was pronounced "dull" during subcommittee hearings, but the proposal failed to excite interest among congressional staff and was simply forgotten when the final version of the bill was drafted.

Require Mandatory Personal Insulation

Local service stations would have been required to provide low-interest loans to individuals desiring to have their blood streams weatherized, had this proposal been successful. All infants born after July 1, 1984, would have been required to be so weatherized at birth, thus guaranteeing them winters without frostbite and summers without heat prostration should their future homes or places of work be without heat or air conditioning. This was to be considered a temporary measure until a cold-blooded human could be developed under programs to be sponsored by the National Science Foundation. The proposal was abandoned when the NSF argued that the National Institutes of Health should sponsor such research, but the NIH felt it should be the responsibility of the Department of Defense. Proponents are reported now to be looking for private funding.

Increase the Speed of Light

Under present physical law light can travel no faster than 186,000 miles per second, and thus it takes sunlight over eight minutes to reach our earth. If the speed limit of light could be raised to, say,

255,000 miles per second, then the sun's energy could reach us in just over six minutes, or about twenty-five percent faster. This would mean the sun's rays would not cool off so much during their multimillion-mile journey, and this attracted the support of solar energy advocates. Furthermore, the energy of matter, which is given by the famous formula $E = mc^2$, would be increased substantially. This brilliant proposal failed to awaken the support of late-sleeping congressmen who feared the greater speed of light would make dawn break earlier than it now does.

Relax the Rules of Arithmetic

The present rules of addition are very rigid, and it is always necessary that $1 + 1 = 2$. One unsuccessful provision of the energy bill would have relaxed the rules so that $1 + 1 = 3$ when counting energy supplies and $1 + 1 = 1$ when counting demands. A group of congressmen firmly opposed to such legislation threatened to filibuster by counting to infinity by ones in the traditional manner should this proposal have come to the floor, because, they argued, such rules would have made vote counting very confusing.

Raise the Absolute Zero of Temperature

For as long as anyone can remember, absolute zero, the lowest attainable temperature, has been about −460° Fahrenheit, at which temperature an electrical current can travel through a wire with virtually no resistance. Most legislators agreed that a warmer absolute zero would all but eliminate the significant energy losses currently experienced in high-voltage transmission lines, but there was a great deal of disagreement over what the new absolute zero should be. It was feared that too high a value would ruin the winter recreation business in Northern states, while too low a value would be of no benefit to Southern states. Thus progress of this proposal through committee was glacial.

Remove the Prohibition Against Water Running Uphill

It is not known when the traditional ban against water running uphill was first imposed, but some legislators felt it was time to repeal the prohibition. Congressmen from mountainous states were tired of seeing their neighboring states generate cheap hydroelectric power from spring runoffs, and they would have liked to have seen some of the runoff coming their way. Congressmen from plains states were flatly opposed to this proposal, however, and they ridiculed the idea of locating hydroelectric dams on watersheds and turbines on mountain tops. During the long deliberations, interest in this proposal evaporated.

Require the Forced Mixing of Oil and Water

Several representatives from coastal states tacked on a proposal to end the de facto segregation of oil and water. Under current practice, the transportation of immiscible oil is subject to stringent environmental requirements that hamper its importation and increase its price. If there were forced mixing of the two substances, oil spills would disperse throughout the volume of the oceans and be less of an immediate threat to the environment, argued proponents. Representatives from inland states had no concrete objections to this proposal, but they withheld their support until they were assured the measure would not make rain-slicked streets even slicker.

Extend the Freedom of Information Act

This proposal would have required Mother Nature to reveal her secrets in response to properly phrased questions from scientists and engineers acting in the public interest. It was hoped that such disclosure would have speeded up the technological realization of advanced energy concepts like fusion, and that it might have led to the accidental uncovering of evidence of energy sources yet undreamed of. This and all the preceding proposals were summarily opposed by a group of congressmen who gave as their only reason their respect for Mothernaturehood and mathematical pi.

23 · MX Decoded

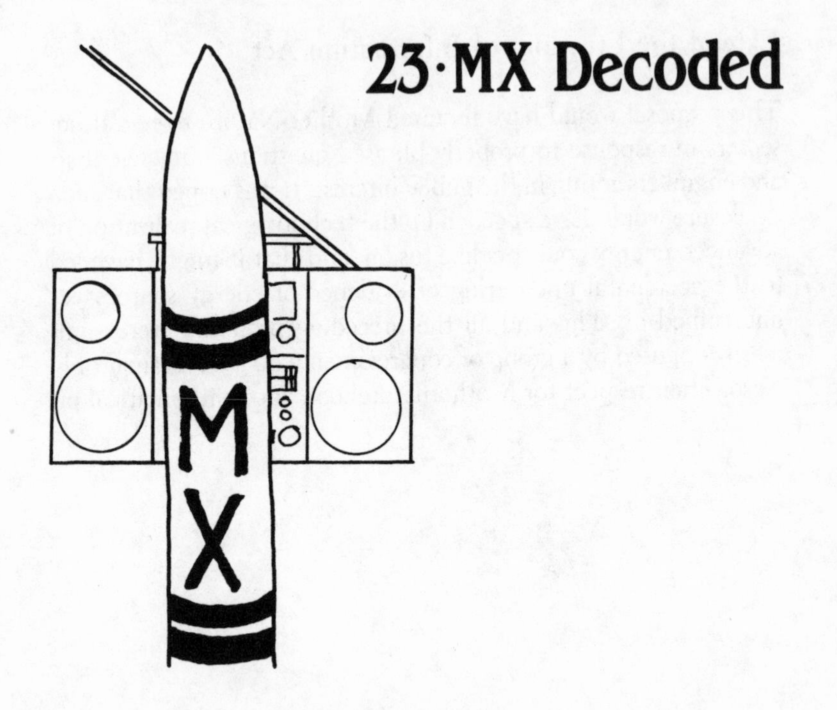

The letters *MX* mean more than just "Missile Experimental," according to anonymous but well-placed Pentagon paper shredders. The letters also stand for the code name of a surefire basing plan. Conceived at the same time as the missile itself, this plan was considered so secret that everyone who was privy to it was immediately debriefed and required to leave government service. The former military strategists dispersed throughout the defense industry, and no one was left in Washington to tell the president or Congress what to do with the MX missiles.

Now the president's commission to evaluate options for deployment is really trying to decode the cryptic message that is *MX*. Among the basing-plan solutions to this puzzle appear to be the following:

• *MX* is really help-wanted-ad jargon for *Male, Experienced.* This suggests that the undeployed missiles be placed among the ranks of the unemployed. The nuclear weapons would then be lost among the millions of Americans who want to work but cannot find a job. They would move about the country looking for positions, standing in unemployment lines too long to be totally destroyed even in an all-out enemy attack.

• *MX* stands for *Mexico.* The idea of this plan is to appear to base the missiles south of the border and then truck them into the States under cover of darkness. Since illegal aliens are notoriously difficult to locate, the missiles should then be successfully based on American soil.

• *MX* stands for *Montreal Expos.* This is the code name for a plan to put missiles under home plate in domestic major league baseball parks. (The Canadian reference is a false signal.) Since arms strategists assume the enemy will play by the rules when attacking us, incoming Russian missiles would be expected to hit first, second, and third bases before they could touch home plate and knock out our capability for retaliation. This would give the defense time to throw the unwelcome visiting team out.

• *MX* stands for *Miter Unknown.* This plan seems to involve the miniaturization of the MX to the point where it could fit under a bishop's hat. Since there are a lot more bishops than missiles, and since they move around a lot, the exact location of the bishops with missiles would be unknown to the enemy. However, the bishops would know, and it is believed that when they found out that they would be asked to wear warheads under their caps they became morally outraged and called for a nuclear freeze.

• *MX* is the ticker tape symbol for *Measurex Corporation* stock. The symbol sounds as if it is traded on the Amex, but that is to confuse the enemy. Measurex is in fact traded on the New York and the Pacific Stock Exchanges, and its use as a code name suggests that the missiles were to be traded as common stock. Predicting where a stock will be on the big board tomorrow is more difficult than finding a missile on a racetrack under the Nevada desert.

• *MX* is really the abbreviation *Mx.*, which bisexuals might prefer to either *Mr.* or *Ms.* If MX missiles were kept in closets with plenty of clothes, no one would know what they were going to look like from day to day. And certainly the Russians are not going to take a missile dressed like "M*A*S*H" 's Corporal Klinger very seriously.

• *MX* is really the Roman numeral for 1010, which is the broadcast frequency of radio station WAR. This is one of those stations that changes its format from hard rock to all talk to no news to good news so capriciously that no one can recognize it from week to week. The basing plan suggested by this code name would disguise MX missiles as WAR radio transmitters, which would only sound as if they were changing locations, thus saving considerable funds to erect still other broadcast missiles.

Critics claim that MX is really nothing but the military-industrial complex giving itself a year-long Merry Xmas and that *MX* really stands for *military excess.* The critics feel the missile xerography that it represents is a major expense the nation can ill afford. Furthermore, they think the endless search for basing plans, while making excuses for those that do not work out, is nothing but a mindless exercise.

24 · Letters to Santa

Dear Mr. Claus:

Christmas Eve has traditionally been a night of confusion for air traffic controllers, not to mention television weather forecasters. This confusion appears to be due in large part to repeated sightings of airborne sleighs pulled by reindeer, and you are believed to be the operator of such an aircraft.

Since it is the responsibility of the Federal Aviation Administration to ensure the safety of the skies by allocating the use of air space, we wish to inform you that, by not filing annual flight plans with the FAA, you have been in violation of federal regulations. If you do not comply with the requirement this year, our enforcement bureau will have no choice but to forbid you to enter our nation's airspace this Christmas Eve.

Thank you for your anticipated compliance.

Administratively yours,
Ebenezer Scrooge

Dear Mr. Cause:

Every Christmas the Environmental Protection Agency listens to countless citizens recite complaints of: a clatter on their lawns, the prancing and pawing of hoofs on their roofs, annoying belly laughs in their living rooms, shrill whistles outside their windows, and, in the early morning hours, exclamations of "Happy Christmas to all, and to all a good night!" out of nowhere. These excessive noises, which disturb the peace on earth of those who have just settled their brains for a long winter's nap, are also reported to be accompanied by ashes and soot swirling from fireplaces and obnoxious pipe-tobacco smoke filling living rooms.

These assaults on the home environment will no longer be tolerated by this Agency, and you are being issued a restraining order (enclosed) against visiting private homes this year.

The order will be in force until you can equip your reindeer with rubber shoes and can demonstrate that you can work in silence and use the front door. Furthermore, you must agree to abstain from smoking in houses without ashtrays.

If you wish to apply for recision of this order, please note that this office will close at 3:45 P.M. on December 24 to wish its staff a happy holiday.

Protectively yours,
Ralph Nader-Rader

Dear Mr. St. Nicholas-Claus:

Two separate complaints, one from a group of young female reindeer at the North Pole and one from an elderly milk cow in Wisconsin, have caused the Equal Employment Opportunity Commission to publish in today's *Federal Register* the following interrogatories to you:

1. What is the sex of each of the following employees of your delivery service: Dasher, Dancer, Prancer, Vixen, Comet, Cupid, Donner, Blitzen, and Rudolph?

2. If, as alleged, all of the above-named employees are male, what affirmative-action programs do you have to add female reindeer to your company's team?
3. Have any of your employees ever jumped over an extraterrestrial body?
4. If they have not, why has a cow, who has jumped over the moon (as witnessed by a cat, fiddle, dog, dish, and spoon) been denied employment as a sleigh-puller?

We shall look forward to responding to your responses.

Equally yours,
Gloria Steinemesis

Dear Mr. Claws:

It has been brought to the attention of the Occupational Safety and Health Administration that you operate a toy manufactory that employs a good number of elves, who, because of their size, must stand on chairs to reach their work. Since this practice presents a clear hazard of falling, you are requested to show cause why you should not: (1) lower the work surface so that chairs are unnecessary and/or (2) equip each chair with a safety railing extending at least 240 degrees around the chair.

The Administration also understands that you deliver your finished products in an open sleigh. Does or will your sleigh have a roll bar?

Have a safe and healthy holiday.

Unreasonably yours,
F.X. O'Sha

Dear Mr. Claes:

The Nuclear Regulatory Commission understands that one of your employees, Rudolph "The Red-Nosed" Reindeer, has a very shiny nose. We are further advised that, if we ever saw it, we would even say it glows.

Since there is a remote chance that substances in this animal's olfactory organ are subatomically active and, therefore, subject to regulation by this Agency, you are requested to answer the following:

1. What is the level of radiation emitted from Rudolph's nose?
2. How many millirems of this radiation will penetrate the average household roof?
3. Are the radiation exposures of the other reindeer and the sleigh operator within regulatory guidelines?
4. What is the total energy release and subsequent radiation exposure that would result from the maximum credible sneeze that Rudolph could release should he accidentally catch cold?

Please respond to these questions within your next half-life.

Regularly yours,
A. Morrie Lovins

Dear Mr. Claas:

Shortly after last Christmas several young people complained to the Federal Trade Commission that you or your elves had used deceptive gift-wrapping practices to conceal the true nature of presents you left at their homes. One boy related the sad tale of unwrapping what looked to be the baseball bat he had asked for on your knee only to find a yellow umbrella. A girl who was sure her soft package from Santa contained a cuddly stuffed hedgehog was disappointed to unwrap a foam rubber football.

Such deceptive practices are contrary to all the truth-in-packaging guidelines of this Commission, and you are hereby served notice to show cause why you should not be restrained from further Christmas giving or from being required to wrap all your gifts in clear plastic.

I remain your humble, civil servant,

Honestly,
David Stockmanacles

Dear Mr. Clause:

It has come to the attention of the Bureau of Food, part of the Food and Drug Administration, that you are somewhat of an unorthodox philanthropist, who has the hang-up of leaving edible gifts in children's stockings. This practice is in violation of several FDA guidelines, and we hope you will clean up your act.

Firstly, food (including candy) can only be distributed if it is packaged in a closed, sanitary container. Used children's socks (often with holes at both ends) do not meet this requirement.

Secondly, any goodies you distribute must have their ingredients and nutritional value clearly stated on a label. If you must leave sweets in (new, unworn, and sealed) stockings, we suggest that you have adequate labels printed to affix to each garment so employed.

Thank you for your generous compliance with our regulations.

Bitterly,
T. H. E. Grinch

25 · Politic Prosody

> *. . . it is advantageous to have a handsome name and one that is easy to pronounce and retain, for thereby kings and grandees recognize us more easily and are less apt to forget us; and even with our own servants, we more ordinarily call on and employ those whose names come easiest to the tongue.*
>
> —Montaigne, *Of Names*

> *. . . the ear is pained by an irregular sequence of air waves which strike the tympanum without any fixed order.*
>
> —Galileo, *Dialogues Concerning Two New Sciences*

What to call himself has assuredly been one of the most difficult aspects of Senator Kennedy's quadrennial decision to become or not to become a presidential candidate. It is not that he is a Kennedy, for that is an asset. It is his first name on which everything might depend. Whether he appears on a ballot as Edward, Edward M., Teddy, or Ted can, as most astute politicians know, but seldom acknowledge publicly, make the difference between victory and defeat.

In the 1976 presidential campaign Candidate Carter insisted that the ballot list him as Jimmy—not James or Jim—Carter, and many political analysts believe that that insistence was pivotal. Had James Carter entered the primaries, he might not have won a one. Had Jim Carter run against Jerry Ford, we might have had an uninterrupted string of Republican administrations.

How can this be? What's in a name? Will not a candidate by any other name run the same? Yes and no, say the poets. Although a rose by any other name may smell as sweet, a political candidate with a prosaic name will run a stinking race. It all has to do with meter and rhyme.

The marketing industry in this country has long recognized the importance of the names of the products it wants to sell. Every poetic device, including onomatopoeia, assonance, alliteration, and even the pun has been used in conjunction with rhyme and meter to sell us what we do not need and may not even want. Before 1976 few Americans thought much about whether they would like a peanut farmer in the White House, but James Earl Carter wanted to be there. So he packaged himself as "Jimmy Carter, peanut farmer," to squeeze the most out of meter and rhyme, conducted a media campaign, and got folks to nominate and elect him. Does that sound so different from the way people are led to select one breakfast cereal over another?

It certainly does not, and longtime marketing consultants, who have helped sell everything from soft drinks to widgets, have become increasingly sought after by politicians. One of the first such consultants to specialize in the verbal packaging of politicians is Dr. Walker Trotter, whose prediction of the outcome of the 1976 presidential election on the sole basis of the sounds of the candidates' names and their campaign slogans and buttons attracted the immediate attention of campaign managers everywhere.

The offices of Preference Meters, the all-for-profit public relations firm that Dr. Trotter heads, are located in Research Triangle Park, North Carolina, and are identified only by the sans-serif letters PM on the front door. Recently Dr. Trotter granted a long-promised interview, and he spoke openly about his theories, for he does not believe the American electorate will ever remember anything it might read even a day or two before any election.

It turns out that PM, which Dr. Trotter founded when he was a graduate student in nearby Durham, is devoted exclusively to advising candidates for high public office in the use of politic prosody, including politic meter and politic rhyme, to their advantage. Politic prosody, according to Dr. Trotter, identifies, classifies, and analyzes the persuasive qualities of our language. Thus, as one of its principal services, PM will scan and subject to a general prosodiacal analysis campaign slogans, promises, buttons, posters, ballots, bumper stickers, and, perhaps most importantly, a candidate's name itself. In short, PM will offer advice on how, in Dr. Trotter's vocabulary, to optimize the aural impact of all aspects of a political campaign.

Dr. Trotter elaborated upon some familiar examples of the effective use of politic prosody to analyze past presidential campaigns. The "I Like Ike" button incorporated the powerful elements of assonance and rhyme in an unbeatable combination. "Nixon's the One" cleverly used rhyme to distract the ear of the electorate from the negative connotations of the first syllable of the candidate's name. "Tippecanoe and Tyler Too" and "All the Way with LBJ" were also very efficacious uses of rhyme. Even John Kennedy, who, like Ted, could not turn meter to his advantage, used the rhyming monogram JFK to woo the ears of the electorate.

Dr. Trotter's theories are based on his dissertation, a definitive study of the meters of American presidents' names, including a comparative scansion of the ballots of our country's forty-seven presidential elections prior to its Bicentennial. Meter—the way the syllables of words and phrases (and names) form patterns of stress and relaxation—is not usually thought of as the province of politicians. It is poets who specialize in using meter's subtle manipulation to get the most out of their words. For meter establishes a rhythmic pattern in a poem, and with its aid a good poet can lead the reader effortlessly through lines he may barely understand, or care to.

But the poets do not have an exclusive license. The scansion of names, according to Dr. Trotter's thesis, involves recording their metrical equivalents, distinguishing the syllables of each name according to whether they are stressed (represented by the solidus, /) or unstressed (represented by the hyphen, –). In his dissertation, Dr. Trotter recognized gradations of stress and caesuras between given names and surnames, but he assured us that a popular exposition of his theories would not suffer from the omission of such refinements. Thus the name Walker Trotter, like Jimmy Carter, is, in the poet's jargon, a double trochee, a regular meter of four syllables in this order: stressed, unstressed, stressed, unstressed. Or, in the notation of scansionists: /–/–.

Such metric regularity, no matter of how short duration, has a strong positive effect on the ear of the electorate, and Americans have had no fewer than fifteen presidents (about one out of every three) with names that scan the same as Jimmy Carter's. Listen to their winning names: Andrew Jackson, Millard Fillmore, James Buchanan, Chester Arthur, Grover Cleveland, Woodrow Wilson, Warren Harding, Calvin Coolidge, Herbert Hoover, Harry Truman, Lyndon Johnson, Richard Nixon, Jimmy Carter, Ronald Reagan.

The list of double-trochee presidents reads like a found poem, and no other single metrical pattern has anywhere near this number of chief executives. While not many students of the presidency will consider this a distinguished list, that itself argues for the power of the double trochee. It is a rhythm with a definite presidential timbre, timbre enough to shore up a weak platform.

The sound of a name, Dr. Trotter holds, strikes the ear first as a metric missile, and only after this has impacted the eardrum does a name travel to the brain as a group of sounds distinguishing the name of one individual from that of another with the same nominal meter. The more striking and satisfying the meter of one's name, the better first impression one will have on the auditor. As to what meters are "better," Dr. Trotter acknowledged that this was a controversial aspect of his thesis. The "better" meters in Dr. Trotter's opinion are those that have—through use, familiarity, and achievement—gained the respect and acceptance, albeit often unconscious, of the great majority of the people.

Preferred nominal meters, such as the double trochees of Jimmy Carter and Walker Trotter, are assets because they possess a regular, repeated pattern and thereby establish an element of trust and predictability in the mind of the hearer. The structure of the first name is echoed in the second, and the pattern that the first name establishes is carried through in the second. Thus the ear is not let down or tricked metrically by the surname. Hence the name Jimmy Carter evoked a confidence that James Carter might not have. The same cannot be said of names complicated metrically to the point where they appear to make fools of the auditor's ears. If one expects, say, a second trochee and does not hear it, one's ears are disappointed, much the way one's outstretched hand is when a practical joker, having extended his hand as if to shake, jerks it back, thumb extended, when we offer ours. A name such as Walker Trot or Gerald Ford has this disadvantage, because the trochee of the first name is not carried through the second. Such disappointments are translated to repressed hostilities toward the person perpetrating the joke, who is of course identified as the person bearing the name, Dr. Trotter maintains.

It is politically advantageous, we were told further, to have a name that begins with a stressed syllable, for this gives rise to a meter that strikes the ear with an initial force, drives deep, and sticks like a barbed spear or harpoon. Moreover, when a barb, or solidus, is followed by a shaft, or hyphen, the ear, and thereby the mind, feels that the name has penetrated deep into the mind and is easily retained—and forgotten with difficulty. Thus /–/– is a better nominal meter than /–/ because it drives deeper into the psyche. Like the flèche on a Gothic cathedral, which anchors a down-to-earth church in the firmament of heaven, the double trochee has a symbolic strength that suggests stick-to-itiveness.

In his dissertation, Dr. Trotter observed that, of all presidents of the United States, only one, Ulysses Grant (–/–/), did not have a barb at the point of his name. It was no surprise to Dr. Trotter that Grant's first opponent's name, Horatio Seymour (–/––/–), not only lacked the initial barb, but was also less regular metrically than Grant's. Dr. Trotter also believes that only Grant's incumbency and military image enabled him to overcome the sound of his name and defeat the metrically very favorable Horace Greeley (/–/–) in 1872. Family names of the presidents follow the same pattern. Of all our presidents, only four, Monroe, Van Buren, Buchanan, and McKinley, lacked initial barbs in their surnames.

The prosodist pointed out that the frustrated 1968 candidate Eugene McCarthy's name, for example, has several of the metrical disadvantages of which he had been speaking. Not only is the name Eugene metrically ambiguous at best, sometimes getting the trochaic pronunciation *Eu*gene (/–), more frequently getting the weaker iambic pronunciation Eu*gene* (–/), but also the surname McCarthy lacks the desirable initial barb. PM's advice to the senator was to use the appellation Gene McCarthy (/–/–) exclusively in any national campaigns, but the advice was not followed.

Dr. Trotter also presented us with some more cold facts from past campaigns. There had been prior to 1976 only four presidential contests involving major party candidates with the meters of Jimmy Carter (/–/–) and Gerald Ford (/–/), and the double trochee had been victorious three out of those four times: Andrew Jackson beat Henry Clay in 1832, Woodrow Wilson defeated William Taft in 1912, Herbert Hoover was victorious over Alfred Smith in 1928, and only in 1908, when William Bryan lost to William Taft, did the double trochee (but a weak double trochee at best) fail to carry the vote. On the other foot, Ronald Reagan meters beat James Carter meters in 1924 when Calvin Coolidge defeated John Davis, and they were victorious in 1892 and 1948 over third-party candidates with names sharing the James Carter meter. Only in 1880, when Winfield Hancock lost to James Garfield, was the double trochee defeated in this way.

The Republicans, who apparently did not consider the metric question in 1976, thus erred when they nominated Ford over Reagan, especially in the wake of Watergate (/–/), to which Gerald Ford was metrically linked. Even Jerry Ford (/–/), the apparent defense against Carter's use of Jimmy, offered no metrical escape. Finally, the Republicans and Ford erred further when they failed to balance a metrically weak ticket with a double trochee. Robert Dole is the metrical equivalent of Gerald Ford, and they might as well have been political twins.

The Democrats, on the other hand, reinforced the metrically favorable Jimmy Carter with Walter Mondale, another double trochee, who, because of the metrical strength of his name, could also have been from Georgia as far as the electorate's ear was concerned. Thus, when Jimmy Carter ran for president in 1976 as Jimmy rather than the apparently more dignified, but metrically inferior, James Carter, he had a distinct advantage for capturing the ear of the electorate.

The 1980 campaign was of course a battle of double trochees, and, all other things being equal, the incumbent double–double-trochee ticket of Jimmy Carter and Walter Mondale should have defeated Ronald Reagan and the metrically brusque George Bush. However, the alliteration of presidential candidate Ronald Reagan carried a lot of aural impact, and everyone knows that more American voters prefer reruns of old movies with the flubs edited out to reruns of incumbent administrations with the flubs on the evening news.

Dr. Trotter also spoke of the 1984 campaign, in which the Democrats again put their best metrical foot forward in the double trochee of Walter Mondale, but stumbled with the metrically unfamiliar running mate Geraldine Ferraro (/–––/–). There is no precedent for a U.S. president with that kind of name, and thus the Democratic ticket seems to have been aurally doomed from the nominating convention.

So this has been Senator Kennedy's dilemma: there is no form of his name that makes him the metric equal of the come-out-of-nowhere Jimmy Carter or the come-out-of-Hollywood Ronald Reagan or the next double trochee that is likely to challenge him for the Democratic nomination for the presidency itself. And with the proliferation of consultants like Dr. Walker Trotter, there are sure to be plenty of double-trochee hopefuls surfacing prior to 1988.

For all the political tradition associated with the Kennedy surname, the stress on its antepenultimate syllable precludes a double-trochaic form that might offset any adverse connotations. Senator Kennedy's only hope lies in the poet's other device: rhyme. So Senator Kennedy's best bet would appear to be to call himself Teddy Kennedy on the ballot. The slant rhyme between the first and last names distracts the ear from the name's inferior meter. Also, Teddy Kennedy invokes the names of two previous popular presidents, and such associations seldom hurt. All in all and measure for measure, while there is some rhyme in every form of the senator's name, Teddy Kennedy emphasizes it most, and that is how someday he should run.

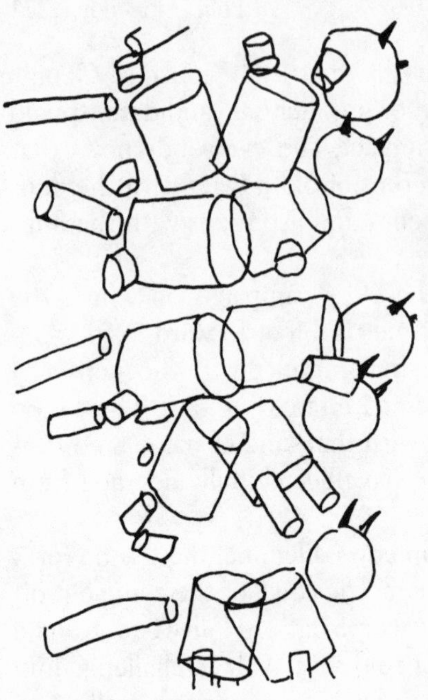

26 · Modeling the Cat Falling

1.

The cat acquires a velocity
At thirty-two feet per second per second
And everything begins to blur—the tail
Is twirling, the cat is turning, the paws

Are on the ground. Again the cat has turned
From upside-down around to downside-down
Without a wall to push on or a string
To pull itself around on the way down.

What is the mechanism? What is the
Solution to the problem of the cat
Released from rest and oriented up,
Descending in a circle in a line

Of gravity, the quickest thinker, down,
To always land with four paws on the ground?

2.
In Comptes Rendus, *in 1894,*
The cat stop falls in photographs Monsieur
Marey has taken with his camera.
His explanation is a hopeless use

Of words for pictures: everyone can see
The cat superimposed upon itself:
The first slow feet, and then the faster feet
Prepared to meet the sidewalk half way down.

Monsieur Marey contorts its vertebrae
With words like torsion, opposition, tors
And wraps its tail around its helix spine
Once clockwise for a counterclockwise half

Rotation of the animal. We see
Two human hands still grasping for the cat.

3.
Herr Magnus (no computer) had to crank
The torso of the cat manually
Through its maneuvers. So he simplified
His model for the numbers it would use.

Dr. McDonald, physiologist,
Dealt with the phenomenological
Aspects. His cat was not an equation
That did its business neatly on the paper.

Professors Kane and Scher were fortunate.
A man on the moon smiled down on them
And NASA sectioned cat cadavers and
Extracted moments of inertia that

The calculus of variations might
Ideally model falling cats in flight.

4.
A couple of cylinders is their cat,
Without a head, with negligible legs,
Without a tail: a mechanical Manx.

This cat possesses a Lagrangian,
Potential and kinetic energy
Confused in an expression for the cat

Falling, the cat jumping, the cat at rest.
The lithe Lagrangian, ready to be
The cat in the clutches of gravity

Submits to differentiation with
Respect to time and with respect to speed
To fall in a falling and revolving mode.

Released now, the equations, upside down,
Descend in the computer, and they turn.

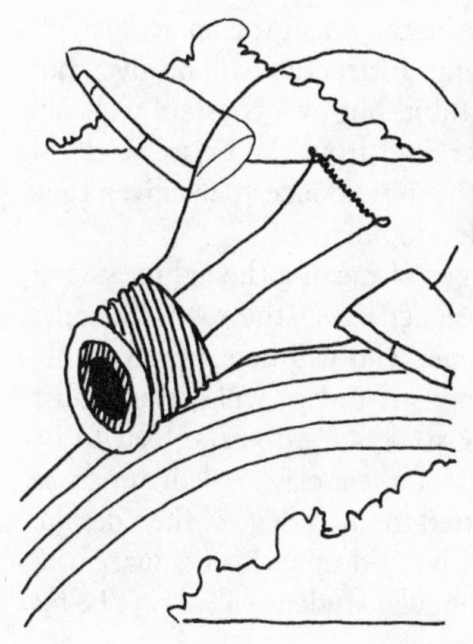

27 · These Goods Better Be Best

Human ingenuity is constantly developing new products and services that extend our reach and make accessible achievements of which only our more eccentric ancestors dreamed. Among some of the most recent additions to the already bulky catalogs of merchandise, services, and concepts to aid us in our pursuits of excellence and excess are these:

A Very Private School

Singular Schools, an exclusive educational consulting firm located in Riverdale, New York, has developed the ultimate concept in private schools. For those parents who find all existing schools inadequate for their only child, Singular Schools will design, have built, staff, and operate for you a unique campus anywhere in the world. Your exceptional child will constitute the entire student body of a school for which he or she can select the name, mascot, song, school colors, and daily menu. Instruction will be given not by just one governess or private tutor, but by a constantly changing faculty of great teachers recruited from the entire world of education. A perfect faculty will be in residence at any given time during the child's intellectual development.

Should the lone pupil show signs of missing the lighter side of education that comes when a teacher leaves the room, Singular Schools can provide a Class Clone, who will be expelled before graduation, thus insuring the singular student will graduate first in his class. Finally, all singular students are guaranteed to be captain of their team, president of their class, valedictorian at their commencement, and admitted to the college of their design. The child's singular campus will be sold upon his graduation to become a seminary, so that the singular student will always be his school's most distinguished alumnus.

New Frontiers

Have you climbed Everest? Landed on the moon? Explored the last frontier?

If you are among the growing crowd of modern adventurers who have accomplished feats like these and are looking for new challenges, you will be interested in the services of New Moons Unlimited, a unique architect-engineer firm headquartered in Antarctica.

The business of New Moons is to create new challenges for idle record-holders. The company has already designed and launched an artificial satellite with an orbit erratic enough to challenge the navigational capabilities of astronauts who have landed on the moon. It has created major new land masses in the South Pacific for veteran nautical explorers. And it has erected mountains steeper and higher than Everest for depressed mountain climbers.

Among the clients of New Moons is the noted sea explorer Jack Questeau, who recently commissioned an enormous lake to be created—in a place unrevealed to him—and stocked with countless species of legendary creatures for him to pursue and photograph. New Moons rose to the challenge and created a crystal-clear lake in the middle of the Sahara Desert.

When asked why he funded a forty-billion-dollar project whose location was unknown to him, Questeau thoughtfully picked up a fistful of water and let it dribble through his fingers. After a few minutes' reflection he responded, "Because it wasn't there."

Two-Wristed Action

A compact digital computer terminal has been developed by California Chronics, a San Diego firm that claims to have perfected the first wrist radio for Dick Tracy. The new device, which looks very much like a fashionable digital watch/calculator, is a full-function computer terminal that enables the wearer to operate interactively in a time-sharing mode any full-size computer within a hundred-mile radius. The terminal has a completely miniaturized alphanumeric keyboard that is operated with the gold point of a five-carat-diamond stickpin, supplied at no extra cost.

Calchron, as the firm is known, also offers a complete line of compatible hardware that may be worn on the other wrist. Among the peripheral equipment is a wrist printer that delivers hard copy looking much like it has come from a Chinese fortune cookie, and a miniature cathode-ray screen that can display fifteen lines per second and has full graphics capability. Wrist disk drives using shirt-button size disks are expected to be announced shortly.

The miniature terminal is available at most full-service gas stations and is being marketed under the name Wristful Thinking.

Condominium Yacht

Why own a fifty-foot cruiser when you can own a stateroom on a 750-foot, 25,000-ton ocean-going vessel? The *S.S. Condominium,* a luxury cruise ship with Liberian registry, Scandinavian officers, and Italian staff, has been sailing from New York to all major ports since its maiden voyage in 1968. Now the operators of this beautiful transatlantic liner are offering discriminating sailors a unique opportunity to own a piece of the boat.

Instead of swabbing your own deck, caulking your own hull, and getting your own ice at the dock, you can relax in one of the *Condominium*'s five hundred deck chairs and pass the world by on your way to the West Indies, Bermuda, South America, Africa, and the Mediterranean. The ship has tow lines for 150 water skiers and ice-making machines to supply 250 dock parties simultaneously.

An owner's association, comprising all stateroom owners, will determine sailing schedules, menus, movies, and the temperature of the water in the ship's seven pools. While individual staterooms may be decorated according to the tastes of their occupants, there are strict regulations as to what may be displayed in portholes.

Staterooms for investment are still available, ranging from efficiency units on the lower decks to interconnected multiple units on the uppers. Don't be left in its wake: sail over to the *Condominium,* kick its life preservers, and see for yourself.

Available Light

If you have been unsuccessful in your photographic attempts to capture the spirit world due to poor lighting conditions and the prohibition of flash bulbs at séances, you may want to try a new film from the Northman Kodiak Company of Anchorage, Alaska. Kodiak's recently announced Any-X Pan Film has a film speed rating of ASA 50,000 and can be push-processed in your darkroom to virtually any ASA rating, a vast improvement over the present practical limit of ASA 3200.

The manufacturer claims that its film can capture the "choir of heaven and the furniture of earth" under the most adverse conditions. This very fine-grained film is expected to be especially useful for immobilizing the most diaphanous features and capturing the most spontaneous motions of the spirit world of poltergeists, phantoms, specters, and the like.

Since most conventional cameras do not have film-speed settings corresponding to that of Any-X, Kodiak has devised a special dark meter for computing proper shutter speeds. The meter is based on the formula $f = 1/(mc^2)$, which gives the f-stop in terms of the materiality m and credibility c of the subject. A specification sheet giving the values of these parameters for different apertures and spooky backgrounds is provided with each roll of film. Any-X is believed to be especially suited for multiple exposures.

The Right Angle

This year's holiday catalog from Kneeman-Markups, the Texas-based department store chain noted for its extravagant merchandise, will have as a unifying theme the golden section. This classic rule of good taste will be carried through the entire catalog, which itself will have the proportions of a golden rectangle—one whose shorter side is to its longer as its longer is to the sum of the dimensions. This proportion has played an important role in the history of art and architecture, and numerous masterpieces have been painted on canvases that are golden rectangles.

The principal offering in the K-M catalog will be a dream house designed entirely on the principle of the golden section. Every floor, wall, window, and door will be a golden rectangle, and the furnishings and appointments, even down to details like light switch plates and floor tiles, will be consistent in their proportions. The house will be constructed on a golden section of prime flatland on the Euclidean Plane of Illinois and will be embedded in dynamic rectangles of corn and soybeans extending as far as the eye can see.

The catalog will contain a golden rectangle template for those who would like to lay out their own house of divine proportion. You can also use this template to check the proportions of all your worldly possessions and, if they don't measure up to the classical standard, you can order ones that do. The catalog is full of them.

Super Transportation

The latest issue of *Areopagitica,* the left-wing newsweekly of the airline industry, has reported that Transcendental Airlines will soon announce that it will be taking reservations for scheduled flights to Shangri-La, El Dorado, Atlantis, Oz, and other destinations heretofore not served by air or any other form of regularly scheduled public transportation. The announcement follows shortly behind that of the Boing Lighter-than-Aircraft Company revealing a major breakthrough in super-optic flight. This development has been expected ever since the first commercial supersonic transport made its maiden flight heard round the world.

Competition had been kept in the dark about the sight-mechanics and aero-optics of the new aircraft, but informed sources now report that the super-optic prototype is shaped like a large incandescent bulb—with its base trailing in the flight configuration—and seats twin copilots across the filament. The commercial model now on the drafting boards is expected to be capable of carrying any prime number of lightweight passengers on the charter flight over the rainbow to the fountain of youth and to have them back before they leave.

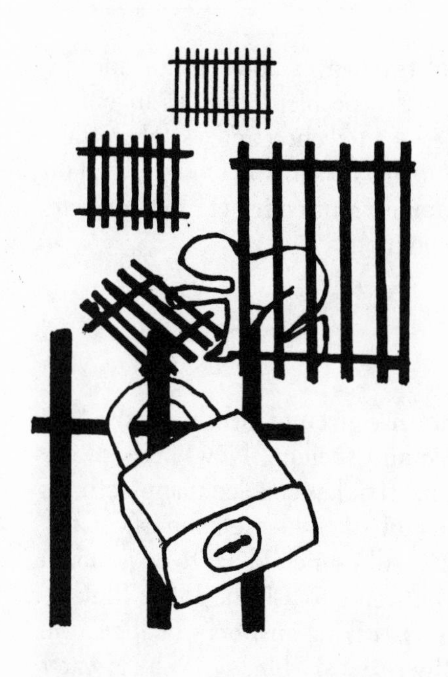

28 · Toys for Parents

Shopping for toys is a depressing experience for many parents. Each Christmas season offers more of the same: toys the parents have to assemble, toys the parents have to play with, toys the parents have to fix, and toys the parents have to pick up.

One of these years, the toy manufacturers will sympathize with the tired parents and offer them something new: toys that give the adults a respite from the constant demands of their children. Some of these dream toys might include the following:

Toy Compactor

Parents would like their children to put playthings away in a toy chest, but the bulkiness of modern toys quickly fills the chest to capacity. Then new toys, which the child is constantly getting, just lie around the room broken and unused after a short while, because the toy box full of old toys leaves nowhere to hide the new. The toy compactor solves this problem, for this ingenious product combines the principles of a toy chest and a kitchen trash compactor. When the child puts his toys in and closes the lid on this toy chest, a powerful hydraulic ram compacts the contents to a fraction of their original volume.

The Little Philosopher

The age of the pocket calculator has given us electronic toys that drill our children in arithmetic and spelling. Now, because the latest breakthrough in miniaturization has enabled manufacturers to put so many memory circuits into a child-size hand calculator, the Little Philosopher is a reality. All your child has to do is punch in the words of the question he wants to ask the Little Philosopher, and in the wink of an eye satisfying answers will appear on the lighted display panel. "Why is the sky blue?" "Why is water wet?" and "Where did I come from?" are only a few of the hundreds of questions your child can pose to this pre-programmed answer man.

The You Are Very Sleepy Book

Numerous children's books are suitable for bedtime reading, and they gently lead up to the idea that the child should go to sleep when the story is over. However, most of these books do not suggest sleep strongly enough, and the child often asks to be read still another bedtime story. The second story is more likely to put the parent than the child to sleep. *The You Are Very Sleepy Book* is different. Written by a professional hypnotist and illustrated by a commercial artist who has worked on subliminal advertising campaigns, this new book effectively suggests, from first page to last, that the child's eyelids are very heavy and that the child will not ask to be read another book. The illustrations cleverly conceal a watch at the end of a chain that swings slowly back and forth as the pages are turned. When you have finished reading this book, your child—arms extended—will quietly go to the bathroom, brush his teeth, kiss you goodnight, tuck himself into bed, and go to sleep.

Construct-a-Cage

This toy comes complete with enough unbreakable plastic bars and locks for your child to build a cage for himself and several of his friends. A cassette tape contains step-by-step instructions for the construction of twenty-four different cage designs ranging from solitary confinement to four-bunk models ideal for sleep-over parties. The color-coded parts are truly easy enough for a child to assemble, but not to disassemble.

The Little Hermit Outfit

This outfit encourages the child to go off by himself and commune with nature, thus leaving you an afternoon to yourself. Included in the Little Hermit Outfit are a sackcloth, rope belt, and sandals costume, plenty of birdseed, a handsome plastic walking stick, and instructions for getting lost in one's own backyard. The instruction booklet also contains suggestions on what to do as a hermit: watch cloud races, wait for visions of sugarplums, learn to talk with the animals, and listen to the plants grow. This outfit is ideal for the only child.

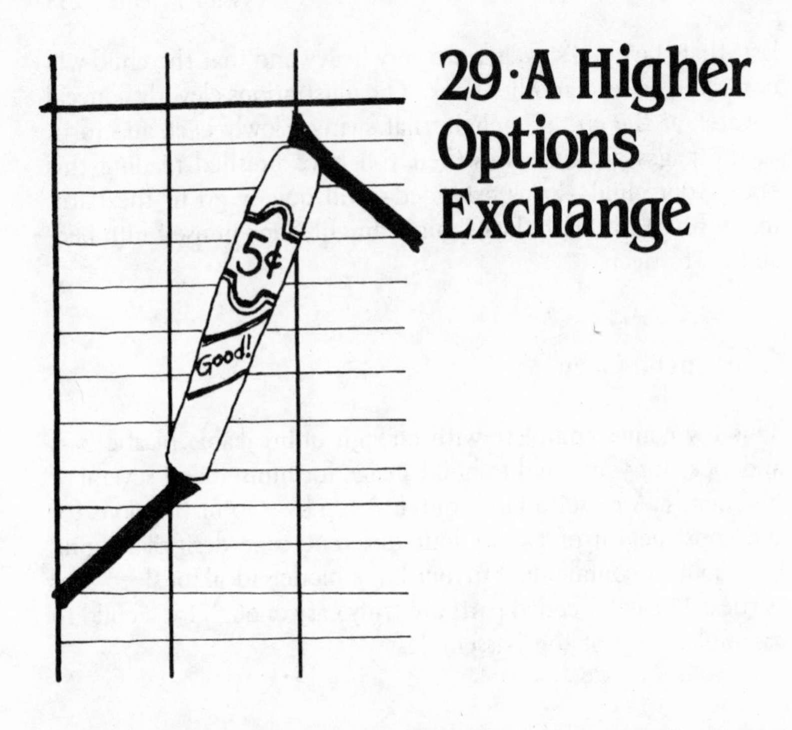

29 · A Higher Options Exchange

Futures and options trading in frozen pork bellies, live hogs, orange juice, and soybean oil has never appealed to a large segment of the investment public. Even common-stock options trading has not attracted the volume of investors its promoters hoped it would. However, there does seem to be a potentially more glamorous options market that may indeed get people to take their money out of certificates of deposit and hand it over to eager stockbrokers.

I propose an options market in which the underlying securities would not be sides of beef or shares of stock but ideas—the fruits of creative thinking. The concept of a Higher Options Exchange, or Hopex, is simple, and its efficacy will be immediately obvious to the Friday Evening Analyst.

This is how the Hopex system would work. Anyone who conceives an idea that he reasonably expects to bring to fruition in a specified period of time could approach one or more brokers on the Hopex and put forth a proposal outlining his idea and the potential financial gain to be realized from it. A broker, having become interested in representing a promising idea, would then submit to the conceiver contracts for exclusive rights to offer a specified number of options in the idea. The contract would typically specify a fixed, one-time fee for the trading right. But particularly promising ideas might command more attractive contracts, especially if there were competition among brokers to represent an idea.

Upon reaching an agreement with the conceiver, a broker would prepare a prospectus describing the idea and the tangible product expected therefrom, the expected date of realization of the idea (which would also be the expiration date of the option), and the financial potential of the idea. The prospectus would also specify the percentage of his royalties that the conceiver is willing to share with an option holder.

Once an idea option was listed, the conventional free market system would take over to set daily prices, which would be reported along with other market transactions in the customary media. A listing for the Hopex might look something like the table on page 236, showing that ideas would be expected to conform to quarterly fruition dates to coincide with the Hopex option expiration dates.

There would be no trading in underlying stocks; the price of an idea option would be affected primarily by rumors of the creator's progress or lack thereof. An inventor's illness would depress the price of an option on his idea because of the anticipated schedule slippage. A report of writer's block would similarly hurt the price of an option on an author's next book.

Conversely, factors that would affect favorably the trading price of an idea option would be an inventor's demonstration of a preliminary working model or a writer's publication of an excit-

Higher Options Exchange
Wednesday, September 7, 1988
Closing prices of all higher options. Sales unit is 100 shares.
a—not traded. b—not offered.

	Oct.		*Jan.*		*Apr.*	
Option	*Vol.*	*Last*	*Vol.*	*Last*	*Vol.*	*Last*
Best Seller	170	11½	b	b	24	7
Better Mousetrap	b	b	a	a	1	½
Bway Hit	75	12½	32	7¼	b	b
Gold Record	16	33⅓	45	78	1	⅞
Good 5¢ Cigar	b	b	7	98	b	b
Great Am Novel	100	¼	12	3	3	12¼
Latest Fad	150	70	200	50½	90	37¼

ing chapter from a work in progress. The relative abundance or scarcity of options in competing ideas would also play a role in setting trading prices.

As an idea option neared its expiration date, the creator himself could purchase his own option if he were confident of realizing his goals—and were able to come up with some cash. Such trading would naturally tend to increase the price of the option. On the other hand, a pessimistic creator, by his lack of interest in his own ideas or his desire to sell any options he himself might own, should depress the price of his idea option.

If, by the expiration date of an idea, the creator does indeed fulfill the promises made in the prospectus for his idea, an option holder could exercise his option, which would entitle him to a share of the royalties on the product of the idea. An option holder might also negotiate at this time with the idea's creator for a one-time cash settlement against any claims to royalties.

Should a creator fail to bring his idea to fruition before the expiration date, the holders of his options would retain the claim to a prorated share of any future royalties realized from the idea. This would prevent the creator from dragging his feet to retain all rights himself.

If an idea option expires without fulfillment, the creator may propose to offer further options on the same idea, to expire at a later date. Any royalties offered in the prospectus would be new royalties—they could not diminish the royalties of the original option holders. While a first prospectus on an idea might offer fifty percent of all royalties to the option buyers, a second prospectus might offer only twenty-five percent, leaving the creator with only twenty-five percent of total royalties. Thus a unique feature of the idea options market would be that expired options could be traded as long as the underlying idea had not been realized and as long as there were still royalties to be offered and realized by the creator.

The creative individual offering an idea option obviously benefits by raising capital with which to realize his idea. This capital would come principally from the negotiated price the broker pays for the right to offer the option on the Hopex. But particularly promising ideas might earn their creators a cut of the broker's commission on all sales that resulted in an increase in the option's price. Creators would never be expected to share in commissions from falling-price sales to prevent the creators from benefiting from their own failures or adverse rumors.

The broker expects to recoup his initial outlay to the conceiver and more, of course, through sales of options alone. The option buyer expects to share in substantial royalties. Potentially, everyone is a winner.

In addition to the idea options listed in the sample closing prices, prospectuses for the following might be expected to be candidates for the Hopex: antigravity machines, philosophers' stones, rhymes for *orange,* substitutes for experience, recipes for free lunches, foolproof schemes, cures for the common cold, ideas whose time will come, and winning market strategies.

30·The Randys: An Immodest Proposal

Unlike movie and television stars, who look forward to the annual Oscar and Emmy award programs that celebrate their industries on prime-time television, scientists and engineers have come to dread the monthly announcements known as the Golden Fleece Awards. These dubious kudos, which often provide comic relief on the evening news, single out for ridicule supposedly frivolous government-sponsored research grants, but every time the Fleece is awarded to a research project with an unintentionally titillating title or to one that treats a subject that could not fail to make school children giggle, the joke is at the expense of all serious scientific and engineering endeavor.

It is time for the technical community to fight back and let the public know that the research-and-development (R-and-D) industry is serious business and not just a wasteful use of federal funds.

An annual gala technical awards program, recognizing on prime-time television outstanding R-and-D achievements and performers, would demonstrate that research and development can both contribute to the advancement of science, technology, and culture, *and* be wholesomely entertaining.

An imitation Nobel Prize ceremony or another of the myriad awards banquets that take place at the specialized technical meetings that only award recipients and their colleagues attend will not do. What the public image of scientific and engineering R and D needs is a televised awards ceremony spectacular enough to rival the Oscar and Emmy shows themselves.

Nominations and voting for the R-and-D awards could be overseen by the National Academies of Sciences and Engineering, and the identities of the winners, sealed in the traditional envelopes, could be among the best-kept secrets of nature. The nominees would be well publicized before awards night, but figuring out the identities of the winners would be beyond any back-of-the-envelope calculation.

Nobel laureates, astronauts, television meteorologists, and scientific and engineering personalities who might frequent late-night television talk shows would host the awards program. Dramatized excerpts from the year's Best Performances of Scientific and Engineering Research and Development would be spread throughout the show, and celebrities would open the sealed envelopes and announce the year's Scientific and Technological Best, as determined by peer review of the nominees.

An adequate number of minor award categories could be established to fill up the early part of the program and build up suspense for the big awards. Among the warm-up awards might be those for: Best Lab Decoration, Best Computer Program, Best Nuclear Reactor, Best ACRONYM, Best High Speed Photography, Best Electron Microscopy, Best Laid Plans of Laboratory Mice and Men, Best Unfunded Proposal, Best Worst Case, Best Bad Actor, Best Effort, Best of Inventions, Best Better Mousetrap, Best Patent, Best Clone, Best Surgical Dresser, Best of All Possible World Lines, Best Equation, Best Fit of Data, Best Blackboard Work, Best Supporting Agency, Best Thin Film, Best Foreign Language Paper, Best Metric Diagram with English Subtitles, and so forth. If the list grew too long, the winners in some of the minor award categories could be announced prior to the televised ceremony.

A rigorous but popular treatment of the mathematics and rules and regulations governing the nominating and voting procedures, including a proof of the existence and uniqueness of the winners, would have to be worked into the awards program. A dynamic young mathematics professor who is into multimedia events and who is used to lecturing to large classes of non-mathematicians would make all this clear to the layman, just as the rules for the Oscars and Emmys are made perfectly clear during those ceremonies.

Since there is absolutely no sex discrimination in the R-and-D industry, there would be no separate Best Female and Best Male Scientist and Engineer. However, the distinctions between pure and applied science and engineering could be expected to be maintained, and whether the Best Engineer or Best Scientist award should be announced first should be a hotly debated issue among the producers of the show. Whatever the penultimate award, the climax of the program would naturally be the Best R-and-D Program award, to be accepted most likely by both the Best Program Director and representatives of the Best Funding Agency.

A suitable statuette, medallion, or trophy would have to be designed, of course, and from this object the awards program and the award itself could be expected to derive their permanent name. Were it not for its pejorative connotations, a natural name for the R-and-D awards would be the Randys. It has even been proposed that the Randy be an androgynous lucite statuette encapsulating a colorful replica of a DNA molecule near its heart and a silicon chip in its brain, but exactly what to call the R-and-D awards and what should symbolize them remains undecided.

There is considerable interest in naming the awards after a famous scientist or engineer. Some suggestions for the totem are: the Leonardo, a multifaceted crystal paperwight depicting prophetic technical drawings from the Florentine's notebooks; the Albert, a terra-cotta bust of Einstein dressed in a sweater and pen and contemplating a piece of chalk; the Isaac, a gold medallion showing Newton standing on the shoulders of some giants and knocking an apple from a tree.

Some feel it is inappropriate to single out one scientist or engineer from the past, and they suggest that the award derive its name from some inanimate symbol of science and technology, such as: the Genie, an empty replica of Aladdin's lamp, to remind us that the spirit of research and development is not in talismans but in the mind of the scientist and engineer; the Astronaut, a statuette of an unidentifiable space traveler in a reflective suit and helmet, to remind us of the many faceless R-and-D technicians behind the scenes of any great project; the C-3PO, an Oscarlike statuette of the tall gleaming robot from the motion picture *Star Wars,* to remind us that the hardest technologists can be soft-spoken and have a sense of humor. The public may be asked, for obvious promotional advantages, to participate in the final selection of a suitable name and symbol.

Whatever the R-and-D awards are named, the sooner they are instituted, the better. It is no secret that foremost among the objectives of public awards programs is to create an enlarged market or audience for the industry sponsoring the awards. The Oscars get more people to the movies, the Emmys increase television audiences, and it is hoped that the R-and-D awards would generate grassroots support for more research and development funding not only by government and private agencies but also by the consumer public.

The gala awards program would almost certainly create scientific and engineering supercelebrities, who would inspire fan clubs. The fans and R-and-D groupies could be expected to converge on technical meetings and hang around the entrances to R-and-D laboratories, hoping to catch a glimpse of the star of the latest awards show and to get an autograph on his latest paper or monograph. As scientists and engineers became more recognizable and celebrated, they would be able to endorse and advertise anything from scientific instruments to breakfast cereals and they would take their place among athletes and actors.

But perhaps most importantly, after the R-and-D awards were firmly established, the Golden Fleece would be shorn of its importance, and the R-and-D industry would be able to praise itself annually and intemperately without distraction.

31 · Metric Sports

The metric movement to replace pounds, quarts, and feet with kilograms, liters, and meters has been hampered by a lack of enthusiasm in this country. This may be due in some measure to the absence of a total commitment to metrication on the part of our great American institutions. Mothers still weigh their babies in pounds and ounces, our flags are measured in stars and stripes, and apple pies are commonly sliced in six to eight pieces. To generate public interest in complete conversion to the more sensible system of multiples of ten, some positive symbols of America, such as baseball and football, must go metric.

The National Metrication Board (NMB), which is responsible for overseeing the nation's conversion to the international system of weights and measures, recently received the long-awaited report from its Subcommittee on Sports. This ten-member body had been assigned the task of trying its hand at rewriting the rules of American sports based on the decimal system of metric measurement, and it has produced a one-thousand-page report weighing a kilogram. The highlights of the document's background reasoning and proposals for metric football and baseball are as follows.

Meterball

The NMB subcommittee recommends that football be renamed "meterball" to emphasize the game's metrication and to remind fans that the foot is passé as a unit of measurement. Meterball would be played on a field one hundred meters long, about ten percent longer than the present playing field. Each meterball team would consist of ten players, and there would be ten officials.

The metric game would be played in ten ten-minute tenths, increasing the playing time by two-thirds. Fans would be forbidden to bring pints or fifths of liquor into the stadium, but in keeping with the spirits of metrication liters would be allowed. The traditional half-time ceremonies would take place after the fifth tenth, and they would be known as the 0.5- or point-five-time ceremonies.

In meterball a team would have ten downs to advance the ball into the end zone. Scoring would be converted to the decimal system with the touchdown being the unit of scoring. But the relative values of other types of scoring would be retained, as follows:

field goal = 0.500 touchdown
safety = 0.333 touchdown
extra point = 0.167 touchdown

Familiar football scores like 14–9 and 31–27 might become 2.334–1.500 and 5.001–3.501, respectively.

To further emphasize the decimal measure of the metric game, the subcommittee has also recommended that such terms as quarterback and halfback be replaced by their decimal metric equivalents, the 2.5-deciback and 5-deciback, respectively.

After the first collegiate metric football game, the "Liter Bowl," was played under a more traditional version of metric rules in Northfield, Minnesota, on September 17, 1977, a spokesperson for the National Football League informed the press that the NFL planned to declare itself exempt from metrication because conversion would be too confusing. In that game, where yards were simply converted to meters, cheers like "First and 9.14, do it again," just didn't bode well for the future of metric football.

But meterball is more than simply metric football, and the Metrication Board now believes that its totally new rules for the game make it less confusing than the simple metric conversion. If the NFL still remains firmly opposed to getting in step with modern systems of measurement, the NMB is prepared to make football a test case for sports metrication by taking legal action against any resisting league.

Base-ten Ball

With present baseball rules, players and fans must memorize such awkward equivalents as:

9 players = 1 side
4 balls = 1 base on balls
4 bases = 1 run
0 runs = 1 shut out
3 strikes = 1 out
3 outs = 1 side out
2 sides out = 1 inning
9 innings = 1 game
162 games = 1 season

This is confusing enough to be included on the back of composition books. It discourages potential new baseball fans and loses them to simpler games like tenpins. The complete and immediate metrication of baseball could help not only the game but also the image of the more sensible system of measurement.

The metrified game, which the recommending committee has called "base-ten ball," would consist of ten-inning games played by ten-player teams. It would take ten balls to make a base on balls, ten strikes to make an out, and ten outs to retire a side.

Since these games could be expected to be about three times as long as a traditional game of baseball, a full season would be reduced to one hundred games and a full team roster would be increased to one hundred players, with the vast majority of them being relief pitchers.

Perhaps the most radical change is the redesigned and standardized playing field that is being proposed to replace the baseball diamond. Base-ten ball would be played on a field one hundred meters square comprising ten bases.

The bases would be arranged in a pattern resembling the outline of the present home plate. Along the foul lines bases will be twenty meters apart, and a circuit around all ten bases will total 250 meters, about two and one-quarter times the present distance a runner must cover to score a run. Since the distance to the first metric base will be about a third shorter than on the present playing field, virtually any ball not caught on a fly will ensure the batter a base hit.

However, it will take more than a conventional double or triple to score a man from first or second base. Since the distance to fifth base is a bit more than the distance around the present base path, a hit that might have been an inside-the-park home run in today's game will be no more than a quintuple or a sextuple in base-ten ball.

Because there are only ten players on a base-ten ball side, it is possible that the nine bases could be loaded with all of a team's players but the batter. In this case the same batter would remain at bat until he made enough outs to retire the side or score the player on ninth base, who then would take his place in the batter's box.

The rules of scoring would be changed to take advantage of the decimal system that is so persuasive a part of metrication. Each player crossing tenth base would score the traditional one run plus a bonus run for his long journey, and players left on base would earn decimal runs. One-tenth of a run would be credited to the score for each base reached. Thus a team that leaves players on first and seventh base would score an additional eight-tenths of a run for that inning. Base-ten ball scores like 27.8–14.3 will eventually become familiar.

The pitcher and catcher would play much as conventional players do now, but the remaining eight players on the team would play new positions. There would be no first baseman because of the near impossibility of throwing a batter out on such a short run, and the pitcher would be expected to cover first and ninth base. A right shortstop would cover second and third base, and a left shortstop would cover seventh and eighth. There would be fourth, fifth, and sixth basemen positioned near those bases, and the three outfielders would also cover third through seventh base. Since they extend quite a bit into the outfield, it is not expected that the pitcher could pick a runner off these bases without allowing a runner on ninth to tag up and score easily. Base stealing would be expected to be frequent, if not exciting, under the new rules, and base runners would be expected to bunch up on the higher-numbered bases.

The cost of converting baseball to the metric system is expected to be in the tens of billions of dollars. Although club owners will have to build new metric stadiums, they will be encouraged to increase their capacities to 100,000 seats. Tickets will cost ten dollars and will bear seat numbers that look like Dewey decimal call numbers. Ten seats will make a row, ten rows a section, and any seat in the ballpark will be as easy to find as a book in the public library.

Notes and References

1. Introduction

The following sources have been quoted:

Samuel C. Florman, *Engineering and the Liberal Arts: A Technologist's Guide to History, Literature, Philosophy, Art, and Music* (New York: McGraw-Hill Book Company, 1968), p. 92.

National Research Council, Committee on the Education and Utilization of the Engineer, *Engineering Education and Practice in the United States: Foundations of Our Techno-Economic Future* (Washington, D.C.: National Academy Press, 1985), p. 73.

Anne Eisenberg, "Mastering On-the-Job Writing Assignments," *Graduating Engineer,* March 1984, pp. 25–28.

David R. Lampe, "Engineer's Invisible Activity: Writing," *Technology Review,* April 1983, pp. 73–74.

John Gardner, *On Becoming a Novelist* (New York: Harper & Row, 1983), p. 142.

Fritz Leonhardt, *Brücken: Asthetik und Gestaltung/Bridges: Aesthetics and Design* (Cambridge, Mass.: The MIT Press, 1984), pp. 32–34.

Edgar Allen Poe is quoted in:

Walter Allen, compiler and editor, *Writers on Writing* (New York: Dutton Paperback, 1949), p. 64.

According to pre-enrollment figures, the entire Duke Class of 1989 was expected to have mean verbal-math scores of 620–661, while the engineering students in the class scored 632–718 on the same SAT achievement tests.

2. Of Two Libraries

A shorter version of this essay appeared in *Technology Review* (January 1981).

The laboratory is Argonne National Laboratory, which is operated by the University of Chicago and located in the western suburbs of Chicago, but I have exercised poetic license in much of my description of the laboratory and its staff.

Data on Newton's Library come from:

John Harrison, *The Library of Isaac Newton* (Cambridge: Cambridge University Press, 1978).

3. Amory Lovins Woos the Hard Technologists

First published in *Technology Review* (June–July 1980).

Amory Lovins, *Soft Energy Paths: Toward a Durable Peace* (San Francisco: Friends of the Earth International, 1977).

4. Soft Technology Is Hard

First published in *Technology Review* (April 1982).

6. Reflections on a New Engineer's Pad

Portions of this essay first appeared in *Technology Review* (February-March 1981).

The engineer's pad was National Paper Company's Stock No. 42–381, priced at sixty-five cents when the slide rule was still pictured. The Keuffel & Esser advertisement appeared on pages 10 and 11 in the September 1950 issue of *Civil Engineering.*

7. Logon Proceeding

First published in *Science 80* (May–June 1980).

The terminal described is a Texas Instruments "Silent 700" Model 745 Portable Data Terminal. Texas Instruments Manual No. 984024–9701–Rev. B provided the specifications.

My office mate and friend was Lech Mync (1945–1979), to whose memory this piece is dedicated.

8. How Poetry Breeds Reactors

First published in *The South Atlantic Quarterly* (Summer 1983).

International Conference on World Nuclear Energy—Accomplishments and Perspectives (Winter Annual Meeting of the American Nuclear Society, Washington, D.C., November, 1980), *Transactions of the American Nuclear Society,* vol. 35 (1980).

An annual survey of nuclear power plants is published in *Nuclear Engineering International.* The May 1985 issue reports that nuclear as a percentage of electrical generating power is 58.7 percent in France and 13.5 percent in the United States.

9. You Can't Tell an Engineer by His Slide Rule

This comprises parts of pieces that appeared originally in *The Washington Post* (January 21, 1982), *The Christian Science Monitor* (January 27, 1982), and the *Duke University Faculty Newsletter* (April 1982).

National Science Foundation and the Department of Education, *Science and Engineering Education for the 1980s and Beyond* (Washington, D.C.: Government Printing Office, October 1980).

10. A New Generation of Engineers

This is an expanded version of a piece that appeared in *The Chronicle of Higher Education* (March 28, 1984).

For an elaboration on the effects of technology on domestic life, see:

Ruth Schwartz Cowan, *More Work for Mother: The Ironies of Household Technology from the Open Hearth to the Microwave* (New York: Basic Books, 1983).

The problem of the rolling spool can be found in virtually any engineering dynamics textbook. The one I was using is:

J. L. Merriam, *Engineering Mechanics, Volume 2: Dynamics* (New York: John Wiley & Sons, 1978). See problems 6/25 and 7/74.

11. The Quiet Radicals

Jerry Rubin is quoted from his letter to the editor, "A 1960s Rebel Looks to the 1980s," which appeared in *The New York Times* for August 25, 1980.

12. Apolitical Science and Asocial Engineering

The bulk of this appeared in *Issues in Engineering—Journal of Professional Activities, Proceedings of the American Society of Civil Engineers* (April 1982).

The epigraph is from:

Herbert Hoover, *The Memoirs of Herbert Hoover: Years of Adventure, 1874–1920* (New York: Macmillan, 1951), p. 133.

The data on occupations of senators and congressmen are taken from the *Congressional Quarterly,* which publishes characteristics of each new Congress.

13. Numeracy and Literacy

First published in *The Virginia Quarterly Review* (Spring 1985).

James B. Conant, *On Understanding Science: An Historical Approach* (New Haven: Yale University Press, 1947), p. 3.

J. Bronowski, "The Educated Man in 1984," *Science*, vol. 123 (April 27, 1956), pp. 710–712.

C. P. Snow, "The Two Cultures," *The New Statesman and Nation*, vol. 52, new series (October 6, 1956), pp. 413–414.

["an expert who must necessarily remain anonymous"], "New Minds for the New World," *The New Statesman and Nation*, vol. 52, new series (September 8, 1956), pp. 279–282.

J. B. Priestley, "Thoughts on Dr. Leavis," *The New Statesman and Nation*, vol. 52, new series (November 10, 1956), pp. 579–580.

C. P. Snow, *The Two Cultures: And a Second Look* (New York: New American Library, Mentor Books, 1964). [I was taken totally by surprise when several young assistant professors—of both cultures—wrote to me asking for a reference to Snow's "two cultures lecture" when I referred to it in an article in *The Chronicle of Higher Education* in 1984. How easily we forget that a younger generation does not necessarily share all of our cultural touchstones.]

F. R. Leavis, *Two Cultures? The Significance of C. P. Snow* (New York: Pantheon Books, 1963).

James D. Koerner, ed., *The New Liberal Arts: An Exchange of Views* (New York: The Alfred P. Sloan Foundation, 1981).

14. A Diary

The year was 1983, but it could be any September. Had it been another week, trouble with my laboratory's equipment, grading examinations, writing a paper, attending a technical meeting, or a variety of other activities might have figured prominently in the schedule.

15. Time Piece

The watches and clocks described are in the 1984–1985 Best Products and Brendle's catalogs.

The story about tasting time comes from:

Daniel J. Boorstin, *The Discoverers* (New York: Vintage Books, 1985), pp. 35–36.

16. Dust Jacket Dilemmas

A shorter version of this appeared in *Prairie Schooner* (Winter 1979–80).

The epigraph is taken from the April 1985 issue of the *Duke University Library Newsletter*, which reprints an item from the *Journal of Academic Librarianship* (May 1983) reporting on the recent discovery of a Mark Twain letter in Massachusetts.

John Carter, *ABC for Book-Collectors*, fourth edition, revised (New York: Alfred A. Knopf, 1966).

17. A Little Learning

First published in *The Reading Teacher* (March 1981).

18. Bullish on Baseball Cards

First published in *The New York Times*, Business and Finance section (July 12, 1981).

19. Outlets for Everyone

This piece was written at the suggestion of Georgann Eubanks when she was editor of *The Guide*, a "lifestyle magazine" for the Research Triangle area of North Carolina, and appeared in the November 1983 issue of that magazine under the pseudonym "Mr. Chips."

The lines of poetry are from Wallace Stevens's "The House Was Quiet and the World Was Calm," which is contained in *The Collected Poems of Wallace Stevens* (New York: Alfred A. Knopf, 1968).

20. Is There a Big Brother?

Approximately the first third of this piece appeared in *The Futurist* (August 1982).

The original "Yes, Virginia" Christmas editorial was entitled "Is There a Santa Claus?" and appeared in the *New York Sun* in 1897.

21. How to Balance a Budget

First published in *The New York Times,* Op-Ed page (April 19, 1980).

22. Washington Entropy

First published in *The New York Times,* Op-Ed page (December 22, 1977) and, in an augmented form, in *Next* (September 1980).

24. Letters to Santa

First published in *The Washington Post Magazine* (December 14, 1980).

25. Politic Prosody

This is a combination of pieces that appeared in *Politics Today* (January–February 1980) and *College English* (September 1980).

The epigraphs are from:

The Complete Essays of Montaigne, vol. 1, translated by Donald M. Frame (Garden City, N. Y.: Doubleday & Company, Anchor Books, 1960), p. 278.

Galileo Galilei, *Dialogs Concerning Two New Sciences,* translated by Henry Crew and Alfonso de Salvio (New York: Dover Publications, c. 1954), p. 108.

26. Modeling the Cat Falling

First published in *Creative Computing* (May–June 1978).

A guest seminar given by Professor Thomas Kane in about 1974 at the University of Texas at Austin led to the following references:

M. Marey, "Des mouvements que certains animaux exécutent pour retomber sur leurs pieds, lorsqu'ils sont précipités d'un lieu élevé," *Comptes Rendus Hebdomadaires des Séances de l'Académie des Sciences, Paris,* vol. 119 (1894), pp. 714–717.

R. Magnus, "Wie sich die fallende Katze in der Luft umdreht," *Archives Néerlandaises de Physiologie de l'Homme et des Animaux,* vol. 7 (1922), pp. 218–222.

D. A. McDonald, "How *does* a falling cat turn over?" *St. Bartholomew's Hospital Journal,* vol. 56 (1955), pp. 254–258.

T. R. Kane and M. P. Scher, "A dynamical explanation of the falling cat phenomenon," *International Journal of Solids and Structures,* vol. 5 (1969), pp. 663–670.

27. These Goods Better Be Best

This piece was commissioned in 1978 by Molly McKaughan, who was then managing editor of *Quest.* Changes at the magazine occurred before the piece could be published in *Quest,* however.

As life imitates art, so the cover story of the June 1985 issue of *Creative Computing* describes a Seiko Datagraph System that is not unlike the wrist terminal imagined here.

28. Toys for Parents

First published in *The New York Times,* Op-Ed page (December 13, 1978).

29. A Higher Options Exchange

First published in *The New York Times,* Business and Finance section (June 4, 1978).

31. Metric Sports

"Meterball" first appeared in *Science 81* (January–February 1981).